Technologies of InSecurity

Technologies of InSecurity examines how general social and political concerns about terrorism, crime, migration and globalisation are translated into concrete practices of surveillance and securitisation of everyday life.

Who are we afraid of in a globalising world? How are issues of safety and security constructed and addressed by various local actors and embodied in a variety of surveillance systems? Examining how various forms of contemporary insecurity are translated into, and reduced to, issues of surveillance and social control, this book explores a variety of practical and cultural aspects of technological control. It also looks at the discourses about safety and security surrounding them. Exploring the inherent duality and dialectics between our striving for security and the simultaneous production of insecurity, *Technologies of InSecurity* considers how mundane objects and activities are becoming bearers of risks which need to be neutralised. Ordinary arenas – such as the workplace, the city centre, the football stadium, the airport and the internet – are becoming imbued with various notions of risk and danger and subject to changing public attitudes and sensibilities.

The book is based on contributions from an international panel of leading criminologists, lawyers and surveillance scholars and provides important new insights about how broader political issues are translated into concrete and local practices of social control and exclusion.

Katja Franko Aas is Assistant Professor at the Department of Criminology and Sociology of Law, University of Oslo.

Helene Oppen Gundhus is Assistant Professor at the Norwegian Police University College.

Heidi Mork Lomell is Post-doctoral Research Fellow at the Department of Criminology and Sociology of Law, University of Oslo.

Technologies of InSecurity

The surveillance of everyday life

Edited by
Katja Franko Aas,
Helene Oppen Gundhus and
Heidi Mork Lomell

Routledge-Cavendish
Taylor & Francis Group
a GlassHouse book

First published 2009
by Routledge-Cavendish
2 Park Square, Milton Park, Abingdon, Oxon OX14 4RN

Simultaneously published in the USA and Canada
by Routledge-Cavendish
270 Madison Ave, New York, NY 10016

Routledge-Cavendish is an imprint of the Taylor & Francis Group, an informa business

A GlassHouse book

Typeset in Times by
RefineCatch Limited, Bungay, Suffolk
Printed and bound in Great Britain by
TJI Digital, Padstow, Cornwall

British Library Cataloguing in Publication Data
A catalogue record for this book is available from the British Library

Library of Congress Cataloging-in-Publication Data
Technologies of inSecurity : the surveillance of everyday life / edited by Katja Franko Aas, Helene Oppen Gundhus and Heidi Mork Lomell.
p. cm.
Includes bibliographical references.
1. Internal security. 2. Security systems. 3. Security (Psychology)
I. Aas, Katja Franko. II. Gundhus, Helene Oppen. III. Lomell, Heidi Mork.
HV6419.T43 2008
621.389'28—dc22 2008009468

ISBN13: 978–0–415–46455–0 (hbk)
ISBN10: 0–415–46455–2 (hbk)

ISBN13: 978–0–203–89158–2 (ebk)
ISBN10: 0–203–89158–2 (ebk)

Contents

6

Contributors

Katja Franko Aas is Associate Professor at the Department of Criminology and Sociology of Law, University of Oslo. She has written extensively on the use of information and communication technologies including *Sentencing in the Age of Information: From Faust to Macintosh* (Glasshouse Press, 2005, shared winner of the 2006 SLSA prize). Her most recent book is *Globalization and Crime* (SAGE Publications, 2007).

Johanne Yttri Dahl is a Research Fellow at the Department of Sociology and Political Science at the Norwegian University of Science and Technology. She is working on the project 'For Whom the Bell Curves', carrying out a study on forensic DNA databases and the use of DNA evidence in Norway.

Benjamin Goold is a University Lecturer in Law and a Fellow and Tutor at Somerville College, and a member of the Oxford University Centre for Criminology. His major research interests are in the use of surveillance technology by the police and the relationship between individual privacy rights and the criminal law. He also writes on aspects of the Japanese criminal justice system and is a member of the Oxford University Faculty of Oriental Studies and an Associate Member of the Nissan Institute of Japanese Studies.

Helene Oppen Gundhus, having previously held a post in criminology at the University of Oslo, is currently Assistant Professor at the Norwegian Police University College. Her theoretical and empirical interests lie at the intersections between crime policy, community safety, policing and culture. Examples include *For sikkerhets skyld* ('For the sake of security', 2006) and several articles in journals including *Journal of Scandinavian Studies in Criminology and Crime Prevention*.

Vidar Halvorsen is Associate Professor at the Department of Criminology and Sociology of Law, University of Oslo. He has written extensively on philosophy of science, philosophy of criminal law and punishment, and ethical issues in police work, including *Ethics, Force and Violence in Policing* (PhD thesis, 2001).

Yvonne Jewkes is Professor of Criminology at the University of Leicester, UK. She has published ten books on various aspects of media and crime, cyber-crime, and imprisonment including, most recently, *Crime Online* (Willan, 2007) and *Handbook on Prisons* (Willan, 2007). She is co-editor of *Crime, Media, Culture: An International Journal*, and series editor of the Sage *Key Approaches to Criminology* series.

Richard Jones is Lecturer in Criminology at the School of Law, University of Edinburgh. He has written on various topics relating to new technology, crime control and criminal justice, including surveillance, policing, cyber-crime and punishment. He is currently conducting research on criminological theories of compliance, social control and regulation.

Francisco R. Klauser is RCUK Research Fellow at the Institute of Hazard and Risk Research at Durham University. Having initiated and led several research projects on the socio-spatial implications of CCTV in Switzerland, his current work focuses on security and surveillance issues in the context of airport risk management and mega sport events.

Heidi Mork Lomell is Post-doctoral Research Fellow at the Department of Criminology and Sociology of Law, University of Oslo. Her research interests include policing, technology and human rights. She is currently studying the use of statistics as a governmental technology, and her next project is *Losing the Discourse of Justice?*, funded by The Research Council of Norway's programme Societal Security and Risk. She is a member of the editorial board of *Surveillance Studies Network*.

David Lyon is Director of the Surveillance Project, Queen's Research Chair and Professor of Sociology at Queen's University, Kingston, Ontario. From 2008 to 2010 he holds a Killam Research Fellowship from the Canada Council. His most recent books are *Surveillance Studies: An overview* (Polity Press, 2007) and *Theorizing Surveillance* (ed., Willan, 2006).

Mike Nellis is Professor of Criminal and Community Justice in the Glasgow School of Social Work, University of Strathclyde. He is a former social worker with young offenders, trained at the London School of Economics in 1977/1978 and between 1990 and 2003 was involved in the training of probation officers at the University of Birmingham. He was awarded his PhD from the Institute of Criminology, University of Cambridge in 1991. He has written extensively on the changing nature of the probation service, the promotion of community penalties, the significance of electronic monitoring and the cultural politics of penal reform (including the educational use of prison movies and prisoners' autobiographies). His most recent book (edited with Eric Chui) was *Moving Probation Forward* (Longman, 2003).

Daniel Neyland is a Senior Research Fellow of Said Business School. He works on a broad portfolio of projects focused on issues of governance and accountability (covering RFID, the global textile trade, the movement of electronic waste, the production of vaccines for neglected diseases of the developing world, airports, traffic management, household recycling and CCTV systems). He has published widely, including two books entitled *Privacy, Surveillance and Public Trust* (Palgrave-Macmillan, 2006) and *Organisational Ethnography* (SAGE, 2007) and has two forthcoming books on mundane governance and an edited collection on 'Privacy, Surveillance and Identity' (co-edited with Ben Goold). Daniel also contributes to Science and Technology and Research Methods teaching at the School.

Gavin J.D. Smith is finishing his ESRC-funded PhD at the School of Social Science, University of Aberdeen. His ethnographic study of CCTV operation explores the phenomenological working practices of CCTV operators in an extensive array of monitoring facilities and examines the ontological composition of such settings. Gavin has a particular interest both in metaphysics and social theory, and is particularly keen to further investigate human emotion and its pivotal role in shaping social action.

Majid Yar is Senior Lecturer in Criminology and Director of the Centre for Criminological Research, Keele University, UK. His research interests include internet crime, intellectual property, and technologies of crime control. He is the author of *Cyber-crime and Society* (2006) and co-author of *Criminology: The key concepts* (2008).

Lucia Zedner is Professor of Criminal Justice, Law Faculty, Senior Law Fellow, Corpus Christi College, and Member of the Centre for Criminology at the University of Oxford, and also Conjoint Professor, Faculty of Law, University of New South Wales, Sydney. She has held visiting fellowships in America, Australia, Germany and Israel. She has published widely in the fields of criminal justice and security.

Acknowledgements

This volume presents contributions which grew out of a conference on 'Technologies of (In)Security' held at the University of Oslo in April 2007. The aim of the conference was to provide critical interdisciplinary perspectives in the field of surveillance, crime control and technology studies. It was an attempt to chart the complex landscapes of surveillance and social exclusion in an era when social control is increasingly technologically mediated and articulated across distant boundaries, as well as marked by various notions of danger and insecurity. We gratefully acknowledge the generosity of the Norwegian Research Council, which provided financial support for the conference and the publication of this book. We also want to thank the University of Oslo, Faculty of Law, for hosting the conference, and our colleagues and master students at the Department of Criminology and Sociology of Law for their participation, support and invaluable assistance. Above all we wish to express our thanks to the contributors who made the conference intellectually stimulating and worthwhile and then wrote the papers that make up this volume. It was an exciting event which we hope comes across in the book as well.

The intellectual trajectory that led to this volume was, from the outset, a collective one, and started with the establishment of the 'Crime control and technological culture' project, funded by the Norwegian Research Council's KIM programme. In the period of 2003–2007 this programme provided not only much needed financial support for our research but also a fertile environment for exciting academic activities and, not least, our friendship. This book is a final record of our collaboration and many debts were incurred in the process. In particular we would like to thank Nils Christie, Thomas Mathiesen and Liv Finstad for their academic involvement at various stages of our project. Many thanks are also due to the staff at Routledge–Cavendish, particularly Colin Perrin and Kate Murphy, for their support and assistance in producing this book.

Katja Franko Aas, Helene Oppen Gundhus and Heidi Mork Lomell
Oslo
February 2008

Introduction

Technologies of (in)security

Katja Franko Aas, Helene Oppen Gundhus and Heidi Mork Lomell

This book, as its title reveals,[1] aims to explore the inherent duality and dialectics between our striving for security and the simultaneous production of insecurity which can result from these efforts. Striving for security is an ambivalent project which carries in itself a potential for creating its opposite – a heightened sense of insecurity. This became obvious in a recent Norwegian conflict between airport-security personnel, primarily Securitas guards, and aircraft crew, such as pilots and flight attendants. In March 2007, the general managers of the three major Norwegian air carriers sent a letter to the airport authorities stating their concern about the excessive security checks. They indicated that the 'unfortunate culture developed by the security personnel' (*VG Nett* 17 March 2007) is in fact damaging to the overall air security. At smaller airports, aircraft crew can be subjected to as many as 10–12 screenings per day. Pilots argued that the frequent controls are often experienced as harassment and make them less capable of doing their job, thereby producing insecurity for the passengers. Crucial issues here are trust and suspicion. Pilots, who are entrusted to fly planes with hundreds of passengers, are yet too suspicious to be allowed to take a lunch break without a security and an identity check. As a result of the conflict, it was decided that the private security personnel would take a course in human relations and would be put under intensified surveillance of about 600 CCTV cameras at Oslo airport, thus creating yet another potential conflict in the work place. Now, the security personnel find themselves under suspicion – 'We are being treated as criminals' is their argument – and their union has complained about the development to the Norwegian Data Inspectorate (*VG Nett* 22 March 2007).

This case of resistance to security measures, both from pilots and security personnel, indicates that not all social groups are equally willing to accept that they are put under suspicion, nor do they have equal possibilities of resistance. While most of us have been disciplined to obediently take off

1 For other takes on (in)security see, among others, Bigo (2005) and Andrejevic (2006).

our shoes and 'spread our arms and legs', for certain social groups the suspicion and humiliation are too much to bear. *Technologies of InSecurity* is a cross-disciplinary book which aims to examine how various forms of contemporary (in)security are translated into issues of surveillance and social control, and vice versa. It explores a variety of practical and cultural aspects of technological control, as well as the discourses about safety and security surrounding them. Who are we afraid of in a globalising world? How are issues of safety and security constructed and addressed by various local actors and embodied in a variety of everyday practices? The volume is structured around five main topics: (in)security and terror, (in)secure spaces, (in)secure visibilities, (in)secure virtualities and (in)secure rights. The book's mainly empirical focus on everyday life is furthermore supplemented with normative issues, particularly the tension between security and rights (see Goold's, Halvorsen's and Zedner's contributions).

There exists by now a burgeoning literature on surveillance, and while engaging with this extensive opus, *Technologies of InSecurity* focuses more specifically on the nexus between various uses of surveillant technologies and (in)security. It thereby aims to connect the field of surveillance studies with the growing body of research about the progressive securitisation of contemporary social life and politics. The securitisation approach enables us to 'unpack' the intricate relationships between technology, crime, danger and fear, and to examine the underlying social dynamics behind various surveillance practices. The security discourse is a way of framing political and social questions in logics of fear and suspicion, where social relations are marked by distrust and uncertainty, particularly with regard to certain social groups defined as security threats (Huysmans 2006). The main focus of the book is on how general social and political concerns about terrorism, crime, mobility and globalisation are translated into concrete practices of securitisation of everyday life, and how mundane objects and activities are becoming bearers of risks which need to be neutralised. Ordinary arenas, such as the workplace, the city centre, the football stadium, the airport and the internet, are imbued with risk and danger and are subject to changing public attitudes and sensibilities. A critical deconstruction of the nexus between everyday surveillance and (in)security has a potential to provide important new insights into how broader political issues are translated into concrete and local practices of social control and exclusion.

Traditionally, prevention of crime has loomed large as the justification for new surveillance practices. However, according to several analysts this task is increasingly being taken over by the broader concept of security. As Zedner (2007a: 265) points out:

> Security is less about reacting to, controlling or prosecuting crime than addressing the conditions precedent to it. The logic of security dictates earlier and earlier interventions to reduce opportunity, to target harden

> and to increase surveillance even before the commission of crime is a distant prospect.

In a security-driven world not only criminologists but also surveillance scholars need a new vocabulary with which to describe and tackle the problems posed by the pursuit of security. The striving for security represents a considerable challenge to existing modes of scholarship by 'stretching existing conceptual and methodological resources to the full' (ibid.: 275). It is our hope that this book will address this conceptual and methodological stretching by introducing some of the new vocabulary that is needed in order to grasp the current developments.

SURVEILLANT TECHNOLOGIES: BETWEEN UTOPIA AND DYSTOPIA

The title of this volume reveals, however, not only a shift towards the new climate of (in)security but also a distinct focus on technology. Much of the debate about the relationship between new technologies and their security effects has been polarised. Critics tend to portray the introduction of new technologies as heralding the advent of a dystopian and totalitarian surveillance society. Their supporters, meanwhile, celebrate them as 'silver bullets', offering the possibility of radically reduced levels of crime and more efficient and effective policing. While understandable, this polarisation of the debate is unhelpful. The powers of surveillance technologies seem at times to be equally believed by their 'salesmen' and their critics and their potentials are often taken for granted without further empirical explorations. *Technologies of InSecurity* aims to function as an antidote to this discourse, by accentuating the need for contextualisation, nuance and ambiguity.

Traditionally, surveillance studies have been somewhat cautious about using the term 'technology', partly due to the fear of becoming guilty of technological determinism. Yet we choose consciously to speak of technologies, partly to avoid clichéd notions of surveillance. Surveillance, like security, has become the buzzword of our cultural zeitgeist. The panopticon and Big Brother associations abound in popular discourse and surveillance literature. Yet, at the same time, these metaphors are increasingly becoming a liability to the scholarly field which they have done so much to promote, and there have been several attempts to break out of the 'panopticon straightjacket' (Boyne 2000, Lyon 2006). Haggerty (2006: 23) thus suggests that the panopticon has become oppressive as a metaphor for analysing surveillance inasmuch as it has become synonymous with surveillance itself:

> [T]he panoptic model has become reified, directing scholarly attention to a select subset of attributes of surveillance. In so doing, analysts have

> excluded or neglected a host of other key qualities and processes of surveillance that fall outside of the panoptic framework.

Also Bauman (2000: 54) draws our attention to the limits of the dystopian visions of the panopticon, the Big Brother and the Brave New World that so many surveillance studies have been based upon. These visions see a future that is tightly controlled, 'a world split into managers and the managed, designers and the followers of designs' (ibid.). Orwell and Huxley 'felt that the tragedy of the world was its dogged and uncontrollable progress towards the split between the increasingly powerful and remote controllers and the increasingly powerless and controlled rest' and could not 'visualize a world without controlling towers and controlling desks' (ibid.: 54).

While the panoptic approach undoubtedly carries much weight in the post-9/11 climate, we hope that by using the concept of 'technologies' we can open up the field to alternative and complementary modes of understanding. In that respect we aim to move closer to Michel Foucault's (1988) meaning of the term 'technology'. Foucault's later work was marked by an attempt to outline the workings of power beyond the famous panoptic design. With a characteristic tendency to generalise, he describes:

> four major types of [these] 'technologies,' each a matrix of practical reason: (1) technologies of production, which permit us to produce, transform, or manipulate things; (2) technologies of sign systems, which permit us to use signs, meanings, symbols, or signification; (3) technologies of power, which determine the conduct of individuals and submit them to certain ends or domination, an objectivizing of the subject; (4) technologies of the self, which permit individuals to effect by their own means or with the help of others a certain number of operations on their own bodies and souls, thoughts, conduct, and way of being, so as to transform themselves in order to attain a certain state of happiness, purity, wisdom, perfection, or immortality.
>
> (Foucault 1988: 18)

The term 'technology' in this context obviously applies to a broad social matrix of action which enables humans to modulate their environments, and panoptic technologies are only one subset among several. Acknowledging the overwhelming focus on technologies of domination, Foucault moved on to analyse 'the interaction between oneself and others and in the technologies of individual domination, the history of how an individual acts upon himself, in the technology of self' (ibid.). This approach, frequently referred to as governmentality studies, combines technologies of domination, such as the panopticon, with technologies of the self and can provide a fruitful path towards transcending the panoptic metaphor (see for example Haggerty 2006).

Technologies of (in)security are therefore not only surveillant technologies but also 'technologies of the self' which are fashioning new, technologically mediated, forms of subjectivity (see also Cole 2006). They are not simply about something affecting and controlling the subject from the outside but are constitutive of who we are and several contributions in this book testify to this development. Daniel Neyland's chapter shows how airport surveillance is far more than simply a task of screening and checking people, but is also about educating passengers about their security responsibilities, changing their consciousness and eventually turning them from 'security unready' passengers to 'security ready' passengers. Neyland outlines the intricate process of 'shifting the ordinary into the threatening'. Scissors, letters and bottles thus no longer are scissors, letters and bottles but become transformed into terror threats. This shift in ontological status of mundane objects is by no means easy and requires active participation and responsibilisation of passengers about the insecure nature of their surroundings. It therefore requires that citizens actively take part in the practices of their own surveillance. Similarly, Mike Nellis' chapter reveals that offenders subjected to satellite tracking are not, and cannot be, wholly passive. They are required to interact with the surveillance equipment, act responsibly towards it and 'embrace the demands the technology makes of them'.

The acknowledgement of the role of self-surveillance in surveillance practices has been one of the vital trends in recent scholarly contributions to the field, described by Whitaker (1999) as 'participatory panopticon' and by Lyon (2006) as 'panopticommodity'. Technologies of (in)security are therefore technologies of *interactive* (in)security (Andrejevic 2006). They are about social interaction, communication and play, challenging and transcending traditional dichotomies between the controllers and the controlled, between the watchers and the watched. As Zedner (this volume) observes, 'the top-down structure presumed by the "sur" in surveillance is by no means inevitable or ubiquitous'. Hille Koskela (this volume) argues that it has become increasingly difficult to define what surveillance actually is. Surveillance has been 'hijacked' by the public and is evading definition. Consequently, many previously clear categories are breaking down: the binary moral opposition between good and bad, between the authorities and the public, the controllers and the controlled. Even CCTV – the epitome of Big Brother and panoptic control – is an interactive medium (Lomell, Dahl and Sætnan 2007). Gavin Smith's chapter in this book takes up the task of looking 'inside the panopticon' and examining the interaction between the watchers and the watched. What he finds are CCTV operators who are both empowered, disempowered and re-empowered by the technologies they operate. 'They are watchers, but also *workers*, subjected to not only the same capitalist regimes of domination as any other labourer in late modernity, but also to an emotive duress produced by the very technologies which earn them their living.'

.echnologies of governance

A vital point is, as Lyon (2001: 27) observes, that surveillance not only constrains but also enables social action. Contemporary social ties have been described as socio-technical ties, 'links that are as much technical as they are social' (Lash 2002: 20). The ability of ICTs to not only watch and record but communicate and articulate is transforming the nature of our knowledge and our sociality (Aas 2005). Technologies are an essential aspect of modernity and through the progressive convergence of production, communication, knowledge and surveillant technologies (encompassing all four points described by Foucault above) they are ingrained in some of the central tasks of contemporary governance. Several chapters in this volume reveal how developments in surveillance are essentially related to the transformations of contemporary modes of governance, particularly globalisation (Klauser, this volume), privatisation (Yar, this volume) and the expansion of managerialism (Nellis, this volume).

Several recent contributions have stressed the centrality of technology to the reconfiguration of what one can call the space of governance. Castells' (1996) extensive opus about the network society envisions technological networks as the main new principle of social organisation. The message has been taken up by surveillance scholars who increasingly point out the centrality of networks in contemporary surveillance practices and the emerging, post-panoptic modes of control (Boyne 2000; Bigo 2005; Bogard 2006). In networks there is no panoptic gaze nor are they marked by the panoptic will to correct (Bogard 2006). The large international networks and databases, such as the European Schengen Information System, form the backbone of contemporary transnational policing. ICTs are expanding the scope of governance and shifting the focus from territorialised to deterritorialised surveillance (ibid.). They have thus emerged as a vital aspect in the denationalisation of state sovereignty and important elements in the emerging global and polycentric modes of governance, sometimes described as governance at a distance (Aas 2005) and policing at a distance (Bigo 2000).

Various forms of contemporary mobility are, as John Urry (2007) points out, crucially dependent on expert forms of knowledge, particularly ICTs, which are increasingly self-organising, co-evolving and interdependent. By implication, this reliance on a technological expert systems is vital also when it comes to control of mobility – what might be called 'immobilization strategies' (Aas 2007). Technologies of (in)security are therefore on several levels connected to the insecurities of heightened mobility. They are about tracking risky mobility, either with regard to transnational migratory flows or on national and local levels. Mike Nellis' chapter in this volume examines the latter by looking at satellite tracking of offenders in England and Wales. Nellis points out that as ever more sophisticated technologies for tracking, tracing and pinpointing develop, 'cultures of locatability' are emerging in a number of organisational fields. The powerful symbolic appeal of incessant surveillance – '24/7/365, anywhere in the world' – is gradually displacing traditional, non-technical and more relational modes of supervision of offenders. As such, Nellis argues, satellite tracking has an inherent affinity to managerialism, which puts premium on efficiency, effectiveness and modernisation. The example clearly stresses the importance of acknowledging the role that surveillance technologies play within internal organisational dynamics and the emergence of new, audit-prone modes of governance, or what Nellis terms 'techno-managerialism'. In this context, the surveillant gaze is directed not only at the individual but also at the state itself, by creating conditions of transparency (Aas 2005).

Furthermore, a vital development in the transformation of contemporary modes of governance is privatisation of various tasks previously reserved to the state, including privatisation of social control and surveillance. Developments in surveillance mirror and support the more general trend towards denationalisation of state sovereignty and privatisation of punishment and social control. Majid Yar's chapter shows how crime control has become a driver of the expansion of commercial surveillance on the internet. Due to the great economic value placed upon informational goods, there has been a growing sense of risk and insecurity on the internet, and growing demands to secure such property. There is, as Yar points out, a range of policing services available to potential cyber-victims, depending on their ability and willingness to pay in the form of subscription and membership fees. Cyber-insecurities are creating the 'hyper vigilant citizen' (Haggerty 2007) as the state appears to be unable and unwilling to take up the task of providing security. Surveillance thus becomes a consumer product, either for reasons of enjoyment and play or for reasons of prudence, following the progressive commercialisation and individualisation of contemporary social life.

However, a vital point here is that privatisation of surveillance is not only a result of the state relinquishing its responsibilities to private citizens but also, crucially, a question of commercial actors having vested interests in the provision of surveillance and security. Francisco Klauser's contribution to this volume describes the relationship between the tremendous security efforts surrounding the 2006 football World Cup in Germany, and FIFA's business interests in the event. Like Yar, Klauser outlines the mutually reinforcing relationship between security politics, surveillance and private business interests evident in the emergence of globalised and privatised security partnerships. Mega sports events, such as the Olympics and football championships, are used as test sites for sophisticated high-tech security technologies. Klauser points out that the FIFA World Cup became a magnified version of some central trends in contemporary security politics: urbanisation of security strategies, globalisation of security partnerships, techno-fixation and intense commercialisation of city space. Security issues related to this, temporarily limited, event crucially affected the whole of Germany and continue to leave their trace in the form of CCTV cameras and long-lasting international security collaborations.

Domesticated technologies

An important lesson that several of the chapters in this book demonstrate is that introduction of new (in)security technologies is by no means a 'smooth ride', as it tends to be portrayed by the proponents of these technologies, as well as by their critics. The social landscape in which various technologies are introduced and implemented is far from seamless and free from conflicts. Occupational cultures and the role of workers and staff are important sites of both mediation and resistance to new policy implementations and there is often tension between old and new ways of doing things (Kemshall 2003; O'Malley 2001, 2004; Gundhus 2005). Studying the everyday practices of security, one may find them far less rational and much more 'messy' than expected. Tools are domesticated, shaped and adapted in occupational cultures and practices. As Lie and Sørensen (1996: 17) put it: 'In theory, technology is a standardizing, globalizing, and bureaucratizing effort. In practice, it is always appropriated and re-embedded in a local context when it is put to use.' This perspective has often been overshadowed in surveillance literature. Its intentionally and unintentionally Orwellian undertones tend to focus on the top-down aspects of watching and technology use and on technology's potential rather than on actual practices. However, '[j]ust because a technology has a potential use, such as that of surveillance, does not imply that this is the use to which it will be put' (Rose 1999: 244).

One of the objectives of *Technologies of InSecurity* is therefore to open up the 'black box' of technology and examine the distinctions between the idealised conceptions of technologies in question and their actual uses. Here, the concept of domestication, developed in the field of science and technology studies, can be of great use as it points out that it is not enough to make the technology available to users in order to make the intended difference. Rather, it is the interaction between availability and use that can tell us something about practice (Silverstone and Hirsch 1992, Lie and Sørensen 1996, Grint and Woolgar 1997). Technologies are seldom used in the prescribed manner which is vividly illustrated in Daniel Neyland's contribution to this volume. Neyland reveals how difficult it may be to reorient organisational activities around new categories. CCTV staff who were given a new priority task of looking for potential terrorists and suspicious packages continued to spend a great deal of time focusing on their regular suspects, such as groups of teenagers. Reorienting understanding was hard to achieve because old ontologies are stubborn and routinised. Similarly, Gavin Smith's study shows that CCTVs are not self-running machines. In Smith's contribution the operators are not simply watchers, who can see what they want, but also workers and human agents, who practically and emotionally adapt to, and domesticate, the technologies they operate. The operators' emotional adaptations to the technology can be seen as an attempt at making it meaningful to one's life – a point which in CCTV literature tends to be overshadowed by a focus on function and utility.

When a technology is domesticated, a practice is developed simultaneously with an interpretation and attribution of meaning. Available technology is modified and appropriated in a specific setting. Users make active efforts to shape their lives through creative manipulation of artefacts, symbols and social systems in relation to their practical needs and competencies. This perspective moves our attention from producers of technologies to non-experts using technologies in their daily activities. Hille Koskela's chapter in this volume points out that while surveillance was previously characterised as a top-down process, conducted by the authorities, in contemporary societies people are increasingly participating in the production of control. This integration of surveillance technologies in everyday lives is a socially contextualised activity, influenced by broader power and social relations, everyday struggles and negotiations. Domestication processes are therefore often accompanied by various strategies of resistance, as also revealed by the above case of aircraft personnel. Gundhus' (2005) study of police use of ICT, for example, found that far from being super-efficient risk communicators, police officers formed 'firewalls of resistance' against transformation (see also O'Malley in Kemshall 2003: 144). Much of this resistance was gendered as certain types of ICT use were associated with femininity and routine-based office work, which was considered inferior to 'real', action-oriented, police work. The domestication perspective emphasises the importance of documenting and analysing how organisational, cultural and structural factors influence surveillance practices. It also directs our attention to the less frequently trodden paths in surveillance studies, such as emotions, organisational resistance, failure and dysfunction.

Technologies of social exclusion

Rather than the Orwellian image of surveillance of everyone, several of the chapters in this book show how the classic theme of exclusion of otherness dominates everyday surveillance practices. Contemporary politics of (in)security is a deeply socially stratified phenomenon. The airport example is in this way exceptional. Airports are one of the few 'domains of generalised suspicion' in society (Feeley and Simon 1994: 182), where everyone seems to be subjected to suspicion and control. In line with general findings within the field, our own studies have found categorical suspicion and social exclusion the basis of much of the surveillance practices (Lomell 2004). The exclusionary aspects reveal one of the paradoxes of security strategies. 'Security is posited as a universal good but in fact presumes social exclusion' (Zedner 2003: 166). As David Lyon writes in his chapter: 'The embrace of the state includes and excludes.' According to Lyon, passports and national ID cards historically have not only enabled the state to distinguish between citizens and aliens but have, with their social sorting capacities, also facilitated mass murder and genocide. The striving to create 'legible citizens' and fix their identity is a central aspect of contemporary security strategies and technologies. 'The border is everywhere' (Lyon 2005), embodied in numerous contemporary practices of 'access control' and social sorting. Borders and frontiers can be set up and defended by dividing cities, squares, public and private buildings (Klauser, this volume) and security checkpoints are becoming ubiquitous aspects of contemporary physical environments (Jones, this volume).

The surveillance gaze is not panoptic or all-seeing but pre-selects its objects of control (Lyon 2003; Goold 2004; Lomell 2007; McCahill 2002; Neyland 2006; Norris and Armstrong 1999). The selection or social sorting is often based on categorical, not on behavioural, suspicion. The insecurity of the majority thus gives legitimacy to various exclusionary practices of marginalised populations. For example, the unease that so-called 'regular' shoppers might feel when sharing the public space with battered drug users is the justification for their exclusion (Lomell 2004). These everyday exclusionary practices from public and semi-public spaces are 'at the margins of criminal justice', where 'the distinction between illegal and unpleasant behaviour, crime and nuisance, delinquency and disorderliness is being eroded' (Hudson 2003: 69). Privatisation of security, combined with the privatisation of public space, results in unequal provision and in circuits of inclusion and exclusion (Zedner 2007a: 274). While police 'proprietorship' extends to the entire civil population – the 'included' as well as the 'excluded' (Waddington 1999: 56) – private security sweeps marginalised segments of the community out of privileged spaces occupied by the wealthy, resulting in not only social but also *spatial* exclusion (Kempa, Stenning and Wood 2004: 564–565). As Klauser shows in his chapter, the owner's right to exclude the unwanted from privatized public spaces tends to 'trump' constitutional protection of individual rights.

The security discourse therefore always begs the question: security for whom? 'Whose security do we pursue? . . . Maximising security therefore necessarily has distributive implications' (Zedner 2007b: 258). Technologies of security and insecurity are significantly also technologies of justice and injustice. The politics of surveillance is a matter not merely of personal privacy but also of social justice. As Nellis points out in his chapter, offenders subjected to incessant satellite tracking do not employ the vocabulary of surveillance but rather of social exclusion: 'Offenders experienced exclusion from a place on the ground more vividly and concretely than they experienced being "watched" by eyes in the sky.'

Technologies of InSecurity makes an argument for even stronger interdisciplinary dialogue between criminology and surveillance studies than has been the case until now. The surveillance discourse's affinity with the language of privacy is not always adequate for making visible the full range of consequenses of social sorting. Moreover, several security practices discussed in this book can be seen as hybrids of surveillance and crime prevention. As Jones suggests in his chapter, security checkpoints at airports should not be reduced to surveillance alone since they bear considerable similarity to situational crime prevention. They mark a different kind of regulatory logic than classical surveillance practices and can effectively operate as standalone systems without links to remote data archives. Nevertheless, the consequences for the persons 'sorted out' raise the perennial normative issues of unfair exclusion, unfair targeting and unfair discrimination (Lyon 2007: 116).

Technologies of (dis)trust

Technologies of InSecurity examines the fearful post-9/11 climate, where the ubiquitous airport screening has become the paradigmatic example of our ingenious striving for security. Securitisation is reaching ever new domains of everyday life and has a particularly profound effect on mobility and air travel. The airport example described at the beginning of this chapter shows how daily lives of air passengers, air crews and airport employees are structured by the security discourse and the conditions of distrust it creates. No one is beyond suspicion – not because we all are potentially dangerous, but because no one can be trusted. In many ways technologies of (in)security are not first and foremost technologies of control but technologies of distrust. The book therefore touches upon some of the central themes of late modernity: risk, trust and disembedding of social relations. As Ben Goold points out in his contribution to this volume, the very existence of surveillance implies some basic absence or withdrawal of trust. Surveillance technologies are symbolic of the state's loss of trust in the public. The paradoxical result is, Goold argues, that 'the untrusting state asks the public to trust it as it expands the apparatus of suspicion and surveillance'.

Trust in technology is intrinsically related to the lack of trust in people (Aas 2006). Surveillance institutionalises distrust and, as Adam Crawford (2000: 208) observes, increasingly, the symbols of trust and security take less the human form of a police officer (or, by implication, a pilot) and more the form of the physical representations, such as the CCTV camera and other security fixtures. 'These are "the physical expressions of a social fabric that defends itself" ' (ibid.), inscriptions of distrust into the physical environment. *Technologies of InSecurity* therefore reveals how security questions today tend to be defined as, and reduced to, technical issues, rather than social ones. The belief in technology as a bearer of security is almost daily confirmed by new forms of technological innovation, such as body scanners, biometric solutions and the like (Jones, this volume). However, technologies of (in)security cannot be seen simply as antithetic to trust relations as such. As Steven Nock (1993) suggests, surveillance measures are essential in establishing trust among the growing numbers of anonymous strangers in our societies. What they establish though is a new kind of trust – 'thin' trust, which is more accessible, but also more fragile (Crawford 2000: 209).

Johanne Yttri Dahl's chapter, meanwhile, reveals how technologies have become not only the entrusted bearers of security but, ultimately, also bearers of truth. Dahl describes the power of DNA as legal evidence and argues that DNA technologies produce the type of truth which is hard to resist and contest for the parties involved. On a similar note, Koskela observes how camera footage often creates an impression of reality where images are easily perceived as being beyond error. Dahl's and Koskela's contributions reveal how the idealised perceptions of the powers of technology gloss over the complex processes of interpretation that lie at the heart of DNA and video camera use.

Nevertheless, the contributions in this book reveal not only the growing belief in the security benefits of various technologies but also how security can be a facade hiding more complex causes and justifications. Jones' chapter in this volume suggests that it is important to examine the symbolic qualities that metal detectors and x-ray machines confer on the institution. It may be a 'mistake to assume that the only, or even the primary, purpose of security checkpoints is that of security', rather it is about projecting an image. Similarly, Yvonne Jewkes' chapter argues that denying prisoners access to the internet, although framed as a security issue, may be underpinned by more emotive objections based on atavistic notions of less eligibility and a wish to segregate, separate and silence those defined as 'others'. Political and media debates about denial of access to the internet serve to highlight offenders' 'otherness' and reinforce punitive approaches to governance. Moreover, Jewkes suggests that silenced in these debates are not only offenders but also voices of experts which could point out that the consequence of intensified social exclusion of prisoners may ultimately be greater insecurity for the community.

An important point here is that the logic of securitisation and the belief in the powers of security technologies are not necessarily scientifically founded. They bring up allusions to Spielberg's *Minority Report*, where the pre-cogs represent the psychedelic core of an otherwise extremely high-tech crime control system. Bigo (2006: 52) thus suggests that the intensification of border controls in the aftermath of 9/11 should not be seen primarily as a sign of increased efficiency, but rather as 'a sign of a ritual against fear of the unknown, with fewer and fewer people believing in the ritual and now seeing it as a simulacrum'. A striking feature of the Norwegian airport security controversy is that some of the central government representatives openly acknowledge the limited security impact of the contested surveillance practices (*Din Side* 14 September 2007). The director general of the

Norwegian Civil Aviation Authority recently admitted that security checks may be excessive in some cases, however Norway cannot afford to become a 'dirty country' by bypassing EU security standards (*NRK* 30 October 2007). Again, trust and commercial interests appear to be vital. In order to be a trusted business partner, an airport cannot afford to become a 'dirty airport'. Airports today are governed by international standards and although these standards may not always create more secure environments, there is security in the standards themselves in terms of predictability, planning, uniformity and commercial advantages. Several chapters in this book therefore outline the reinforcing relationship between security and economy, where security presents itself as an economic imperative. However, even though this may seem a straightforward relationship, as Neyland shows, when security interferes with consumerism there can also be too much security.

CONCLUSION

Technologies of InSecurity aims to raise questions about the broader cultural and social implications of various security technologies. Several observers have pointed out the centrality of the striving for security for the maintenance and articulation of contemporary political and social order (Zedner 2007a; Huysmans 2006). The tendency to make security pervasive has become 'a dominant, emotionally charged element of political culture and everyday life' (Loader and Walker 2007: 11), also described as the 'culture of fear' (Furedi 2006). Late-modern societies seem not only to be governing crime, and through crime (Simon 2007), but also governing security and through security (Johnston and Shearing 2003). Bigo (2006: 51) describes the development as the 'governmentality of unease' fuelled, among others, by the proliferation of 'professionals of (in)security'. However, the 'anxiety market' is riddled with paradoxes (Lianos and Douglas 2000: 120; Zedner 2003). The extended use of socio-technical environments produces safety without social interaction, with the social consequences that the 'delivery of safety leads to demands for its increase' (Lianos and Douglas 2000: 121).

In extreme cases, Vidar Halvorsen's chapter reveals, the fearful cultural climate and the notion of society under constant threat can lead to fatal ethical consequences. Halvorsen points out that this is particularly evident in contemporary versions of the so-called 'ticking bomb scenario', which regularly informs the plot of the popular TV series *24*. The bitter war on terror in the series confers a strong sense of urgency, emphasised by the constant ticking of a digital clock not interrupted even by commercial breaks. This pervasive sense of urgency, as Zizek (2006) points out, has profound ethical implications:

> The pressure of events is so overbearing, the stakes are so high, that

> they necessitate a suspension of ordinary ethical concerns. After all, displaying moral qualms when the lives of millions are at stake plays into the hands of the enemy. CTU agents act in a shadowy space outside the law, doing things that 'simply have to be done' in order to save society from the terrorist threat.

This sense of urgency, therefore, not only informs the world of military intelligence gathering and policing but through the incessant bombardment by popular television finds its way to our living rooms on a daily basis. One of the aims of *Technologies of InSecurity* has been to outline the progressive embeddedness and domestication of various technologies in the practices of everyday life. This embeddedness may be part of the explanation of why people generally are willing to accept surveillance – a cause of much exasperation on the part of privacy activists.

A mobile phone has become a central tool for police surveillance (Gundhus 2006), precisely because it has become almost an extension of the body. The top-down surveillance is possible because of the embeddedness of technologies in everyday life – in the mundane aspects of everyday tasks, in the production of knowledge and in the modes of governance. These days, the presence of a surveillance camera on a UK high street may have become so commonplace that people forget that they are being watched and the authorities envision a need for a shouting CCTV (Norris 2007).

The question can be raised, of course, whether, by pointing out the mundane nature and embeddedness of technologies, we create an even more sinister and pessimistic analytic framework, where there is 'no outside' and where everyone is potentially implicated in practices of surveillance. If we all are implicated in surveillant practices – either because we are insecure or because of the technologically mediated nature of our daily tasks – what space is there for traditional modes of resistance, such as, for example, the language of privacy and human rights? When technologies are so pervasive – and surveillance is increasingly coming close to entertainment – we need a new vocabulary of resistance and new modes of regulation which transcend the conventional opposition between public and private, the state and the citizen. As Goold points out in his contribution, this new vocabulary cannot be based only on individualistic notions of privacy and civil liberties. Collective costs to the social fabric (and not least those left outside of it) are just as essential for understanding the risks of surveillance as individual threats to liberty and privacy.

References

Aas, K.F. (2007) *Globalization & Crime*, London: SAGE.

Aas, K.F. (2006) ' "The body does not lie": Identity, risk and trust in technoculture', *Crime Media Culture*, 2: 143–158.

Aas, K.F. (2005) *Sentencing in the Age of Information. From Faust to Macintosh*, London: Glasshouse Press.
Andrejevic, M. (2006) 'Interactive (in)security. The participatory promise of ready.gov', *Cultural Studies*, 20: 441–458.
Bauman, Z. (2000) *Liquid Modernity*, Cambridge: Polity Press.
Bigo, D. (2006) 'Security, exception, ban and surveillance', in D. Lyon (ed.) *Theorizing Surveillance: The panopticon and beyond*, Cullompton, Devon: Willan Publishing.
Bigo, D. (2005) 'Globalised in-security: the field of the professionals of the unease management and the ban-opticon', *Traces: A multilingual series of cultural theory*, 34–87.
Bigo, D. (2000) 'Liaison officers in Europe: New officers in the European security field', in J.W.E. Sheptycki (ed.) *Issues in Transnational Policing*, London: Routledge.
Bogard, W. (2006) 'Surveillance assemblages and lines of flight', in D. Lyon (ed.) *Theorizing Surveillance: The panopticon and beyond*, Cullompton, Devon: Willan Publishing.
Boyne, R. (2000) 'Post-panopticism', *Economy and Society*, 29: 285–307.
Castells, M. (1996) *The Rise of the Network Society*, Oxford: Blackwell Publishers.
Cole, M. (2006) 'The role of confession in reflective practice: Monitored continuing professional development (CPD) in health care and the paradox of professional autonomy', in D. Lyon (ed.) *Theorizing Surveillance: The panopticon and beyond*, Cullompton, Devon: Willan Publishing.
Crawford, A. (2000) 'Situational crime prevention, urban governance and trust relations', in A. Von Hirsch, D. Garland, A. Wakefield (eds) *Ethical and Social Perspectives on Situational Crime Prevention*, Oxford: Hart Publishing.
Din Side (14 September 2007) 'Vil be EU fjerne væskeforbudet'. Available at: http://www.dinside.no/php/art.php?id=396855 (accessed 14 February 2008).
Feeley, M. and Simon, J. (1994) 'Actuarial justice: The emerging new criminal law', in D. Nelken (ed.) *The Futures of Criminology*, London: SAGE Publications.
Foucault, M. (1988) 'Technologies of the self', in L.H. Martin, H. Gutman and P.H. Hutton (eds) *Technologies of the Self: A seminar with Michel Foucault*, London: Tavistock Publications.
Furedi, F. (2006) *Culture of Fear Revisited: Risk-taking and the morality of low expectation*, London: Continuum.
Goold, B.J. (2004) *CCTV and Policing: Public area surveillance and police practices in Britain*, Oxford: Oxford University Press.
Grint, K. and Woolgar, S. (1997) *The Machine at Work: Technology, work and organisation*, Cambridge: Polity Press.
Gundhus, H.O. (2006) *'For sikkerhets skyld'. IKT, kunnskapsarbeid og yrkeskulturer i politiet*, Oslo: University of Oslo.
Gundhus, H.O. (2005) ' "Catching" and "Targeting": Risk-based policing, local culture and gendered practices', *Journal of Scandinavian Studies in Criminology and Crime Prevention*, 6(2): 128–146.
Haggerty, K. (2007) 'Care of the virtual self: identity theft and its victims', paper presented at conference 'Technologies of (In)security' at University of Oslo, April 2007.
Haggerty, K.D. (2006) 'Tear down the walls: On demolishing the panopticon', in D. Lyon, D. (ed.) *Theorizing Surveillance: The panopticon and beyond*, Cullompton, Devon: Willan Publishing.

Hudson, B. (2003) *Justice in the Risk Society: Challenging and re-affirming justice in late modernity*, London: SAGE Publications.

Huysmans, J. (2006) *The Politics of Insecurity: Fear, migration and asylum in the EU*, London: Routledge.

Johnston, L. and Shearing, C. (2003) *Governing Security: Explorations in Policing and Justice*, London: Routledge.

Kempa, M., Stenning, P. and Wood, J. (2004) 'Policing communal spaces', *British Journal of Criminology*, 44: 562–581.

Kemshall, H. (2003) *Understanding Risk in Criminal Justice*, Maidenhead: Open University Press.

Lash, S. (2002) *Critique of Information*, London: SAGE Publications.

Lianos, M. and Douglas, M. (2000) 'Dangerization and the end of deviance: The institutional environment', in D. Garland and R. Sparks (eds) *Criminology and Social Theory*, Oxford: Oxford University Press.

Lie, M. and Sørensen, K.H. (1996) 'Making technology our own? Domesticating technology into everyday life', in M. Lie and K.H. Sørensen (eds) *Making Technology Our Own? Domesticating technology into everyday life*, Oslo: Scandinavian University Press, 1–30.

Loader, I. and Walker, N. (2007) *Civilizing security*, Cambridge: Cambridge University Press.

Lomell, H.M. (2007) *Selektive overblikk. En studie av videoovervåkingspraksis*, Oslo: Universitetsforlaget.

Lomell, H.M. (2004) 'Targeting the unwanted: Video surveillance and categorical exclusion in Oslo, Norway', *Surveillance & Society*, 2: 346–360.

Lomell, H.M., Dahl, J.Y. and Sætnan, A.R. (2007) 'Å se og bli sett: Kommunikative aspekter ved videoovervåking', in N. Levold and H.S. Spilker (eds) *Kommunikasjonssamfunnet. Moral, praksis og digital teknologi*, Oslo: Universitetsforlaget.

Lyon, D. (2007) *Surveillance Studies: An overview*, Cambridge: Polity Press.

Lyon, D. (2005) 'The border is everywhere: ID cards, surveillance and the other', in E. Zureik and M.B. Salter (eds) *Global Surveillance and Policing: Borders, security, identity*, Cullompton, Devon: Willan Publishing.

Lyon, D. (2001) *Surveillance Society: Monitoring everyday life*, Buckingham: Open University Press.

Lyon, D. (ed.) (2006) *Theorizing Surveillance: The panopticon and beyond*, Cullompton: Willan Publishing.

Lyon, D. (ed.) (2003) *Surveillance as Social Sorting: Privacy, risk and digital discrimination*, London: Routledge.

McCahill, M. (2002) *The Surveillance Web: The rise of visual surveillance in an English city*, Cullompton, Devon: Willan Publishing.

Neyland, D. (2006) *Privacy, Surveillance and Public Trust*, Basingstoke: Palgrave Macmillan.

Nock, S.L. (1993) *The Costs of Privacy*, New York: Aldine de Gruyter.

Norris, C. (2007) 'Sound and vision: Some critical reflections on "talking" CCTV', paper presented at conference 'Technologies of (In)security' at University of Oslo, April 2007.

Norris, C. and Armstrong, G. (1999) *The Maximum Surveillance Society: The rise of CCTV*, Oxford: Berg.

NRK (30 October 2007) 'Luftfartsdirektør advarer pilotene'. Available at: http://www.nrk.no/nyheter/distrikt/nordland/1.3887421 (accessed 14 February 2008).

O'Malley, P. (2004) *Risk, Uncertainty and Government*, London: Glasshouse Press.

O'Malley, P. (2001) 'Policing crime risks in the neo-liberal area', in K. Stenson and R.R. Sullivan (eds) *Crime, Risk and Justice: The politics of crime control in liberal democracies*, Cullompton, Devon: Willan Publishing.

Rose, N. (1999) *Powers of Freedom*, Cambridge: Cambridge University Press.

Silverstone, R. and Hirsch, E. (eds) (1992) *Consuming Technologies: Media and information in domestic spaces*, London: Routledge.

Simon, J. (2007) *Governing Through Crime: How the war on crime transformed American democracy and created a culture of fear*, New York: Oxford University Press.

Urry, J. (2007) *Mobilities*, Cambridge: Polity Press.

VG Nett (22 March 2007) 'LFF mener Gardermoen-ansatte blir overvåket på jobb: Blir behandlet som kriminelle'. Available at: http://www.vg.no/nyheter/innenriks/artikkel.php?artid=146632 (accessed 14 February 2008).

VG Nett (17 March 2007) 'Norwegian, SAS Braathens og Widerøe i felles brev: Securitas-opptreden går utover flysikkerheten'. Available at: http://www.vg.no/nyheter/innenriks/artikkel.php?artid=169940 (accessed 14 February 2008).

Waddington, P.A.J. (1999) *Policing Citizens: Authority and rights*, London: UCL Press Ltd.

Whitaker, R. (1999) *The End of Privacy: How total surveillance is becoming a reality*, New York: The New Press.

Zedner, L. (2007a) 'Pre-crime and post-criminology?', *Theoretical Criminology*, 11: 261–281.

Zedner, L. (2007b) 'Seeking security by eroding rights: The side-stepping of due process', in B.J. Goold and L. Lazarus (eds) *Security and Human Rights*, Oxford: Hart Publishing.

Zedner, L. (2003) 'Too much security?', *International Journal of the Sociology of Law*, 31: 155–184.

Zizek, S. (2006) 'Jack Bauer and the ethics of urgency', *Times*, 27 January. Available at: http://www.inthesetimes.com/article/2481/ (accessed 14 February 2008).

Part I

(In)security and terror

Chapter 1

Mundane terror and the threat of everyday objects

Daniel Neyland

INTRODUCTION[1]

The surveillance studies community has grown significantly over the last 10 years and continues to be characterised by lively debate, detailed empirical studies and engagement with a broad range of theoretical perspectives. However, this chapter will argue that objects (rather than people) as matters of concern have hitherto been somewhat neglected. Ordinary and mundane objects form a central focus for the governance, accountability and surveillance relations which characterise our everyday life. This chapter will suggest that objects have become a poignant matter of concern for surveillance activities as a result of particular articulations of terrorist threats. Objects and the possible threats they pose have thus become matter(s) of security and insecurity.

Over several decades the UK has been witness to fluctuating assessments of the likelihood, source, possible targets and mode of operation of terrorism. From IRA attacks through to the July 2005 London bombings, facets of mundane activities have suddenly come under scrutiny in response to apparent terror threats. Getting on a bus or train in London, walking past a rubbish bin or working in a tall office building have quite rapidly become matters of concern. A feature of this concern has been that specific categories of previously ordinary, everyday object have also been called to attention. Backpacks and other luggage (either attached to people or detached), plastic bags, cars and vans, and recently water bottles or other liquid containers have at times shifted from the ordinary, comfortable and everyday into categories of suspicion. These moments of scrutiny for everyday actions, places and things are often accompanied by announcements regarding ways in which the population should manage its relationship with those matters of concern (usually under the terms that populations should be vigilant regarding articulation of

1 Many thanks to the editors and conference participants for their helpful comments on earlier versions of this chapter.

the expected modus operandi of the threat). Furthermore, the same matters of concern are taken up either formally or informally by surveillance systems in, for example, city centres and airports, operationalising these focal points in activities of data collection and scrutiny.

This chapter will analyse the shifts of ordinary and everyday things, actions and places into matters of concern by presenting research from three recent projects (one on urban CCTV, one on airport security, one on counter-terrorist security advice). Each example will be used to analyse the work that goes into shifting the ordinary into the threatening, encouraging the population to take up particular new relations with the things around them, and the consequences of attempts to shift things from the mundane into matters of concern.

OUR RELATIONSHIPS WITH MUNDANE OBJECTS

Mundane ordinary stuff raises challenging, provocative and unsettling questions of our sense of security and insecurity when it goes bad. It is a common motif within horror films for family members, friends and work colleagues to be transformed into flesh-eating zombies, for beloved pets to turn rabid and for objects which we take for granted (TVs, cars, cutlery) to become possessed by evil spirits. What was close and familiar becomes terrifying (this closely aligns with neo-Freudian analyses of what is sometimes termed the 'uncanny'). But what happens when an ordinary object becomes a matter of concern in our everyday lives? When a letter becomes a (potential) bomb, when a water bottle causes airport security anxiety, when a black rubbish bag becomes the focal point for a large-scale police operation designed to capture an extortionist/terrorist/latter-day Robin Hood? And how can social science get to grips with these sudden transformations?

A natural starting point for an analysis of (in)security and everyday objects might appear to be surveillance studies. Here we find detailed and thought-provoking studies of CCTV systems (e.g. Norris and Armstrong 1999), train stations (e.g. Muller and Boos 2004) and airports (e.g. Adey 2004) each implicated in forms of surveillance. And we find detailed analysis of privacy in relation to its legal implications (Goold, this volume) and our own sense of identity (Lyon, this volume). However, within this broad and diverse community of scholars we find little attention paid to the nature, materiality and identity of objects involved in surveillance. The privacy, identity and surveillance of people rather than things are the focus of attention. However, with objects increasingly coming under scrutiny, being identified as a threat to our security and legitimised as a focus for surveillance, how can we get to grips with ordinary things, what role do they play in surveillance and how can a focus on things open up interesting, provocative and challenging questions for surveillance studies?

This research drew together three areas of social-science research to engage with these questions: to engage with mundane and ordinary objects and attempt to get to grips with how ordinariness is accomplished; to understand the ways in which these objects might be situated at the fulcrum of a series of relationships which attempts to establish the identity of the objects and our 'proper' relationship with them; and as an attempt to understand the nature of these relationships, ideas on governance and accountability were explored.

First, this research engaged with Science and Technology Studies (STS) in order to develop an understanding of the roles of ordinary and everyday objects in social interactions. Whereas in previous research STS has often emphasised the need to examine high-profile debates about scientific and technological controversies (e.g. GM crops, BSE, genetic bio-weapons), the emphasis in this research has been on the mundane technologies of everyday life. These latter deserve analytic attention because they tend to have been overlooked in the face of high-profile causes championed by the media (for examples, see Shove and Southerton 2000; Shove 2003; Latour 1992; 1991). Usefully much recent STS proposes that techno-scientific artefacts are central to the genesis and maintenance of networks of social relations. This is in line with the familiar slogans of STS, for example that 'technology is politics by other means', or that 'technology is action at a distance', or that techno-scientific artefacts are 'congealed social relations'. STS also emphasises the need to eschew technical determinism. So, for example, Actor Network Theory (ANT) (Callon 1986; Latour 1991; Law 1991) begins to articulate these themes by showing that the 'effects' or 'consequences' of technological entities can be understood as the upshot of heterogeneous relations of assemblages or networks. For this research, these ideas suggest that shifting a letter into the category (potential) letter bomb could be understood by paying close attention to the ways in which the properties of the (ordinary, everyday) letter are shifted through the establishment of the (threatening, potential object of terror) letter bomb and that this process would be the upshot of a complex assemblage of network relations. The next move for this research would then appear to be to identify and come to some understanding of the network relations which establish and maintain these identities for objects.

However, the precise status and role of the constituent entities remain unclear in these network approaches. Is this no more than a form of 'distributed essentialism' more or less straightforwardly 'fixed' by the network? Can the apparent completeness of networks adequately account for the possibility of unrepresented others (Lee and Brown 1994)? For example, would it make sense to talk of the network around letter bombs as including the letter bomber alongside potential victims and, if so, how would the resulting simultaneous different viewpoints of the letter be incorporated and managed in a single network? These difficulties have led to recent efforts to reconceptualise the nature of the artefact at the heart of relations, for example as a deferred contingency (Rappert 2001; 2003), as a fluid technology (de Laet and Mol

2000) or as a blank figure (Hetherington 1997). Other attempts include the articulation of the role of technical artefacts in social ordering (Law 1994), configuring the user (Grint and Woolgar 1997) and performing social communities (Woolgar 1996). Particularly useful for this research was Mol's (2002) suggestion that multiple ontologies can exist simultaneously. In line with these ideas, the objects in focus for this research could be conceived as occupying several different ontological positions simultaneously. This research explored these suggestive themes specifically in terms of the accountability relations involved and the role of mundane technological artefacts in maintaining (or disrupting) networks of governance.[2]

In order to understand the role of objects as being at the heart of networks of governance and accountability, secondly this research focused on several literatures detailing the social organisation of accounting and accountability relations. This field of research analyses, for example, social contingency in the production and use of accounting systems (Power 1997; Baxter and Chua 2002). If one conceives of, for example, the letter bomb as being a focal point for the development of a governance and accountability network, this contingency would suggest that the network is perhaps more fluid than early ANT would allow. A key argument for these governance approaches is that social control is achieved through forms of discourse (Foucault 1977), calculation (Rose 1996) and categorisation (Bowker and Star 2000) by these sets of relationships. So the network is not limited to establishing identities for objects such as letter bombs, but would also be involved in articulating and establishing relations between those entities connected through the network and how they should treat the object in focus. In this sense, the network of governance and accountability relations would establish the ways in which, for example, letters should be handled given the potential of letter bombs, thus establishing the new identity for the object at the heart of accountability relations (the letter) as a matter of concern (a potential bomb). This would also simultaneously establish identities for others connected in the network as those who should take responsibility for handling letters in particular ways.

When discussed in relation to systems of audit, Foucauldian-inclined analyses suggest that social control or forms of governance occur by virtue of a process of internalisation of categories and values (Miller 1992; Miller and O'Leary 1994; Rose 1999). Similarly, Ericson et al's recent (2003) depiction of 'insurance as governance' offers a highly suggestive picture of a system of categories and identities providing forms of social control in the context of increasing state abrogation of regulatory activity. The disadvantage of these approaches for this research is that they tend not to deal with mundane techno-scientific objects and they sometimes provide scant detail about how

2 Also helpful as background to this research was the (mainly US-based) research on the social construction of public (social) problems (e.g. Bash 1995; Gusfield 1996; Adams 2003; Woolgar and Pawluch 1985).

internalisation works in practice. So although a general feature of this argument might be that, for example, the letter bomb's identity as a letter bomb and the identity of other entities in the network as potential letter bomb handlers is the upshot of members of the network internalising a categorical schema (of letters, bombs, etc.), this does not tell us much about where these categorical schemas come from or how the schemas might be put into practice. Nor do these discourse-oriented approaches pay much attention to the contingent local apprehension, interpretation and uses of the technologies involved. What happens if no one pays attention to letters as potential letter bombs?

In order to get to grips with this apparent fluidity at the centre of networks of governance and accountability, this research drew, thirdly, on the use of accountability as a central concept of ethnomethodological research (Drew and Heritage 1992; Luff et al 2000; Suchman 1993). Here the term accountability usually defines a generic condition for the possibility of social interaction (Garfinkel 1967) and tends to refer to immediate interactions in, say, a conversation (e.g. Hutchby and Woofitt 1998). In this sense accountability derives from those moments where, for example, a first speaker provides an utterance which is held to account by the second speaker in their response and this is made available to the first speaker as a demonstration that what has been said makes sense. Moments of accountability are thus constitutive in that making sense of what is being said in a conversation, for example, operates as a moment of accountability which constitutes the sense of the interaction taking place. Although this appears to be a complex way of talking about something relatively straightforward (a conversation), it usefully establishes a means to recognise the contingency of accountability relations. A network of governance and accountability relations focused around a particular object would not necessarily fix the identities of those entities involved in the network and, instead, the fluidity of their identities becomes a focus for research. A drawback of these ethnomethodological approaches is that it remains unclear where the boundaries for accountability might be drawn (Neyland and Woolgar 2002; Neyland 2006) and how this contingent sense-making might relate to the initial question for this research which aimed to understand the shift of objects into matters of concern.

In sum, this research looked to address the absence of objects in surveillance studies by drawing together insights from STS on objects and networks, Foucauldian-inspired research on ideas of governance and ethnomethodological ideas of contingency in accountability relations. This combination suggests that letter bombs, for example, could be addressed as objects at the centre of networks of governance and accountability relations; that these networks of relations could be central in establishing the nature and identity of the object in focus and the roles and relationships of those entities drawn together into the network; and that these identities, roles and relationships are the contingent upshot of interactions in the network (that is, the relations, identities and objects are all available for further shifts). It appears that the objects at the centre of these networks may experience something like a reconfiguration of their ontological identity with, for example, letters becoming potential bombs (drawing on ideas from the work of Mol 2002). The following sections will now look at examples in practice and ask: how are ontological shifts in the nature of ordinary and everyday objects accomplished? In what ways are these shifts tied into networks of governance and accountability relations? And what are the consequences of these networks and attempts at ontological shifting?

Example 1: letter bombs

This first example comes from a recent piece of research by the author looking at the rise of letter bombs as matters of concern. The particular focal point here will be the website of Britain's intelligence service MI5. Britain's intelligence service has led a varied life, established in 1909 as the Home Section of the Secret Service Bureau. At the outbreak of the First World War in 1914 Britain established various military operation sections. Section 5 of military operations was dedicated to intelligence and gained the name MO5 (Military Operations Section 5). In 1916 these operations went through a change of terminology in order to decrease confusion over what each section actually did in practice. This led to a re-branding of the section as Military Intelligence, section 5 or MI5. Subsequent to the war, the military designations were dropped and MI5 became the Defence Security Service in 1929 and then the Security Service in 1931. However, the name MI5 persists as a routine naming convention for the activities of the intelligence services in the UK (and indeed its website address retains the name: www.mi5.gov.uk). In place of a wartime focus on providing intelligence for the defence of the nation, MI5 now operates under the same forms of bureaucracy as many government offices. MI5 has to demonstrate its transparency (making certain types of information available), value for money (under annual assessments of returns on spending) and its usefulness as a government department. Under the auspices of the latter principle, MI5 presents itself as a useful source of information on the current threat of terror facing Britain, the likely source and objects of terror, and establishes itself as the provider of training on how to manage terror.

One of the significant domestic terror threats that MI5 has highlighted in the last few years has been letter bombs. Its website contains a section dedicated to the threat. Much of the information contained on the website is designed to, first, suggest who should be concerned by letters, secondly, to emphasise the threat posed by letters and, thirdly, to encourage those who should be concerned to adopt particular activities in relation to letters. The MI5 website makes it clear that small to medium-sized businesses are a particular target for letter bombs. It emphasises this risk on the website and also by e-mailing small to medium-sized businesses with the suggestion that employees should be informed of ways to manage this threat. (Indeed, my own university department received this advice which was then subsequently distributed by e-mail to all members of staff.) The instructions to pay attention are as follows.

Letter bombs

- Letter bombs, which include parcels, packages and anything delivered by post or courier, have been a commonly used terrorist device.
- A properly conducted risk assessment should give you a good idea of the likely threat to your organisation and indicate precautions you need to take.
- Letter bombs may be explosive or incendiary (the two most likely kinds), or conceivably chemical, biological or radiological. Anyone receiving a suspicious delivery is unlikely to know which type it is, so procedures should cater for every eventuality.
- A letter bomb will probably have received fairly rough handling in the post and so is unlikely to detonate through being moved, but any attempt at opening it may set it off. Unless delivered by courier, it is unlikely to contain a timing device.

This information is also an initial way of suggesting who should be taking responsibility for managing the risk of letter bombs. Thus much attention is given to the post-rooms and the post-handling staff of small to medium-sized businesses.

This emphasis on post-rooms moves us into the second area of MI5 activity which focuses on separating out the letter bomb as a potential threat. We are told that small to medium-sized businesses are at risk, that post-rooms should be taking responsibility for managing this risk, and then we are told what kinds of signs to look out for so that we might successfully manage this risk. MI5 offers the following as a way of recognising a letter bomb.

Indicators of a letter bomb

- It is unexpected or of unusual origin or from an unfamiliar sender.
- There is no return address or the address cannot be verified.
- It is poorly or inaccurately addressed, e.g. incorrect title, spelt wrongly, title but no name or addressed to an individual no longer with the company.

- The address has been printed unevenly or in an unusual way.
- The writing is in an unfamiliar foreign style.
- There are unusual postmarks or postage paid marks.
- A Jiffy bag, or similar padded envelope, has been used.
- It seems unusually heavy for its size. Most letters weigh up to about 30g, whereas most effective letter bombs weigh 50–100g and are 5mm or more thick.
- It has more than the appropriate value of stamps for its size and weight.
- It is marked 'personal' or 'confidential'.
- It is oddly shaped or lopsided.
- The envelope flap is stuck down completely (a normal letter usually has an ungummed gap of 35mm at the corners).
- There is a pin-sized hole in the envelope or package wrapping.
- There is an unusual smell, including but not restricted to almonds, ammonia or marzipan.
- It has greasy or oily stains on the envelope.
- There is an additional inner envelope and it is tightly taped or tied (however, in some organisations sensitive material is sent in double envelopes as standard procedure).

The websites and e-mails from MI5 separate out those whom might be at threat from letter bombs and those whom should take responsibility for managing the threat and also separates out those items which should become a matter of concern. Through this detail we can begin to see a network of relations being built around the mundane object. Small to medium-sized businesses, post-rooms, MI5 advice, letters, and letters with a particular appearance are drawn together in relations of risk, threat and responsibility.

The third area of information on the MI5 website provides information on appropriate actions by those who come into contact with potential letter bombs.

Although any suspect item should be treated seriously, remember that the great majority will be false alarms and a few may be hoaxes. Try to ensure that your procedures, while effective, are not needlessly disruptive. Take the following into account in your planning:

- Seek advice from your local police CTSA on the threat and on defensive measures.
- Consider processing all incoming mail and deliveries at one point only. This should ideally be off-site or in a separate building, or at least in an area that can easily be isolated and in which deliveries

can be handled without taking them through other parts of the building.

- Make sure that all staff who handle mail are briefed and trained. Include reception staff. Encourage regular correspondents to put their return address on each item.
- Ensure that all sources of incoming mail (e.g. Royal Mail, couriers, hand delivery) are included in your screening process.
- Ideally, post-rooms should have independent air conditioning and alarm systems, as well as scanners and x-ray machines. However, while mail scanners may detect devices for spreading chemical, biological and radiological (CBR) materials (e.g. explosive devices), they will not detect the CBR materials themselves. At present, no CBR detectors are consistently capable of identifying all hazards reliably. Post-rooms should also have their own washing and shower facilities, including soap and detergent.
- Staff need to be aware of the usual pattern of deliveries and to be briefed of unusual deliveries. Train them to open post with letter openers (and with minimum movement), to keep hands away from noses and mouths and always to wash their hands afterwards. Staff should not blow into envelopes or shake them. Packages suspected of containing CBR material should ideally be placed in a double-sealed bag.
- Consider whether staff handling post need protective equipment such as latex gloves and face masks (seek advice from a qualified health and safety expert). Keep overalls and footwear available in case staff need to remove contaminated clothing.
- Make certain that post opening areas can be promptly evacuated. Rehearse evacuation procedures and routes, which should include washing facilities in which contaminated staff could be isolated and treated.
- Prepare signs for display to staff in the event of a suspected or actual attack.

Here we find that not only are objects (in this case letters and packages) to be understood in new ways as part of a new category (as terrorist threats) and not only are particular people at risk (small to medium-sized business post-rooms), but also that those at risk and those who should take responsibility for managing the threat should reorient the organisation's activities around the threat. So a network is built around the object (letters), drawing together various entities in relationships of risk, threat and responsibility (toward the management of letter bombs), those relationships are to reorient the activities of the organisation around the object in focus (letters should be handled in new and different ways) and this establishes the transformed ontology of the object in focus (no longer a 'letter' but a 'potential letter bomb'). This scheme gives us a working basis for beginning to understand how mundane and everyday objects become the focal point of terror and a matter of concern. However, although this first example provides us with rich instructional detail, it ends some way short in terms of the actual practicalities of putting this scheme into practice. First, what are the challenges in reorienting the design of an organisation's activities around an object's new ontology? Second, to what extent are these reorientations picked up by those who are deemed to be responsible for managing the threat posed by the object's new ontology? The next two examples will look at these questions in detail.

Example 2: matters of concern in airports

This second example is drawn from research carried out by the author in a major international airport just outside London through which upwards of 20 million passengers travel each year. The research involved ethnographic study of airports by a team of ethnographers, spending time with airport managers (on tours of the airport and discussing their jobs) and survey-type interviews with 400 airport passengers. This section will focus on the work done by airport managers to engage with the threat posed by the re-designation of passengers' objects as terrorist threats.

During a routine tour of the airport, the terminal manager and terminal technology development manager were keen to express their three principal interests: security, retail and efficient passenger movement. The terminal manager suggested that his airport prided itself on its security record, that passengers travelling from that airport were known around the world (mostly by other airport managers, but also by government departments) to have been checked to the most stringent standards. He referred to passengers who had been security checked as 'clean' and newly arrived passengers as 'dirty'. Attributions regarding the capacity of the airport's security system to successfully and appropriately divide, categorise and assess passengers formed important arguments regarding the identity of the airport and the ability of the managers to do their job. Simultaneous to this security focus, the airport operated a business model emphasising the importance of retail income (it generated more income from shops than from planes taking off and landing). The managers were keen to figure out how they could enhance efficient passenger movement in order to maintain security levels, while reducing time 'wasted' by passengers (in queues, taking off coats for the security checks and so on) and thus increasing the amount of time passengers had access to retail outlets.

While other airport managers formed the focus for claims regarding the integrity of security, the airport's board of directors formed the audience for considering retail income. The airport operated a formula for assessing the number of seconds each passenger spent in the departure lounge shopping area and the annual income this temporal indicator should produce. Thus the capacity to govern passengers in such a way that they moved through security quickly and shopped slowly was a central focus for accountability relations connecting the managers to their board. However, airport managers were also held to account by passenger representatives through the Passenger Services Group. This group looked for improvement in passenger experience in the airport, brought problems with airport wayfinding to the attention of managers and sought to represent particularly important customer types for the success of the airport's business model (such as regular business passengers and high-value holiday users).

For the airport managers, their three principal interests (of security, retail sales and efficient passenger movements) and the relations of accountability into which they entered (with government departments focused on security, their board of directors focused on business models and the Passenger Services Group) were closely intertwined with newly emerging concerns regarding the production of 'security-ready' passengers. The managers were keen to reduce time 'wasted' by passengers turning up at the security check not ready to go straight through the checks. They wanted passengers to have already taken off their coats and got their bags ready to go through the x-ray machines and they wanted passengers to arrive at the security check without any 'sharps' (mostly scissors or knives) and, later, liquid containers larger than 100ml. In order to produce security-ready passengers who were not carrying sharps, airport managers followed a similar set of activities to that identified on the MI5 website. First, the managers sought to identify those with responsibility for managing the threat of sharps (those who should not be trying to carry them on to the plane). The group singled out for attention were departing passengers. Second, the managers sought to separate out and re-categorise those objects which should form the focus of attention (in this case 'sharps', which were mostly knives and scissors). Third, airport managers attempted to reorient passengers' actions and attitudes towards these objects and establish their new ontology. These were no longer to be considered scissors, but were to be understood as terror threats. The new activity required of passengers to affirm this new ontological status was that these objects should not be in passengers' hand luggage presented to airport security.

Did this reorientation work? The first attempt by airport managers to reorient passengers' activity came prior to the August 2006 security alert focused on liquid containers. This attempt to get passengers to reorient their actions around the new ontological status of their ordinary objects involved placing boards at the entry points to airport security, encouraging them to, for example, take off their coats.

These boards provided instructions on the activities passengers needed to carry out in order to be 'security ready' when they arrived at the x-ray machine. As far as the managers could tell, these boards had little impact on passengers' behaviour as many still presented themselves to airport security staff in a 'security-unready' state and there was no noticeable increase in passenger through-put times in the security area and thus no discernable increase in retail income (which the managers hoped would derive from fast security and slow shopping). The managers suggested that perhaps passengers had better things to occupy their time such as talking about forthcoming holidays. In order to reaffirm the ontological status of 'sharps', airport security had to maintain 'knife buckets' into which any sharp items which passengers attempted to carry through security would be placed. 'Sharps' were thus separated physically from their passengers by security staff; this process was slow, passengers did not take as much responsibility for managing the terror threat of scissors as managers wished and the new ontology of scissors as a matter of concern was left uncertain.

Airport managers made a second attempt to reorient passengers' activities around their ordinary and everyday objects by augmenting the boards with plasma screens suspended above queuing passengers. These screens played looped animation of what passengers should be doing while they waited in line for the security check. Once again this appeared to have no discernible impact on passengers, through-put, security check times or what is known in airports as 'retail dwell time'. The managers suggested that perhaps passengers 'don't look up'.

A third attempt to reorient passengers' understanding of their mundane objects and how they should be treated involved the security scare of August 2006 and newly emergent worries over liquid containers.[3] Faced with the prospect of even slower security checks, less retail dwell time and a greater number of ordinary objects to be considered threats (not just 'sharps' now but also bottles of water), a new system was introduced. This involved placing leaflets into the hands of departing passengers as they checked in at the airports.

The research concluded at the same point as this new system was introduced. However, judging by the security chaos that ensued immediately following attempts to transform the ontology of liquid containers from, for example, bottles of water to terrorist threats, it appears that this third system was, like the first two, similarly messy and inconclusive on introduction.

In sum, it appears that the working model produced in example 1 enables us to get to grips with some features of the shifting of objects from ordinary and everyday to matters of concern. In a similar manner to MI5, airport managers attempted to establish who should take responsibility for a particular terror threat, separated out and re-categorised objects as now being a matter of concern and attempted to re-orient people's actions towards those

3 It was suggested during this crisis that there was reliable information from the security services that terrorists were planning on using liquid containers to blow up aeroplanes in flight. It has been a matter of some dispute as to the veracity of this threat.

objects to confirm the new ontological status. However, attempts to get passengers to pay attention to this new ontological status and to reorient their actions in line with this new status were messy. Queues initially remained longer than the managers wanted in airport security and as the number of matters of concern broadened, the queues lengthened. Although it seems reasonable to argue that, for example, bottles of water formed the focal point for the building of a network of relations held together by relations of risk, threat and responsibility, and these relations were to a degree focused on establishing a new ontological identity for bottles of water, this ontological transformation was not straightforward. The number of ways in which passengers did and did not reorient their actions around bottles of water and the numbers of ways in which water was and was not a matter of concern led to confusion and delays in the airport and a profusion of matters of concern (including the impact delays would have on retail dwell time, whether or not this would lead to disgruntled shareholders selling their stakes in the airport operator, whether or not this would threaten the airport managers' jobs and so on[4]). Perhaps this mess of relations and interpretations was down to the nature of airports, with so many passengers coming from different parts of the world and requiring varied instruction on objects' ontology. Example three will look in more detail at the ways in which employees in a less chaotic organisation reorient their actions around newly emerging matters of concern.

Example 3: CCTV and the Mardi Gras bomber

Example 3 is drawn from an ethnographic study of CCTV (reported more extensively in Neyland 2006). This aspect of the research drew on the study of one town in 1997–1998. The Mardi Gras[5] bomber was a UK-based mundane terrorist who carried out a 'reign of terror' between 1994 and 1998 (although his activities were not always deemed terrorism and he was also dubbed a latter-day Robin Hood; see BBC 1998; BBC 1999). At first the bomber targeted a major chain of UK banks with suspicious packages sent through the post (these were mainly in the form of video cassettes designed to explode once opened, although many of them did not explode). The bomber then moved on to target a major chain of UK supermarkets, leaving suspect packages (in the form of black rubbish bags) containing explosives in the car parks. These packages also contained the bomber's calling card.

The bomber attempted to extort money from these organisations by demanding cash in return for a cessation of attacks. The bomber threatened to change tactics if money was not forthcoming, and to follow customers home from the supermarket and shoot them with a cross-bow. The bomber

4 In the end, all of this happened.

5 Sometimes also named the Mardi Gra bomber.

and police communicated through the Lonely Hearts adverts of a national newspaper in the UK.

The bomber had caused minor injuries at this point, but police officers suggested that it was only a matter of time until the bomber killed someone. It was at this point that the author was sat in a CCTV control room in a town just outside London on a Thursday afternoon to evening slot. Not a great deal appeared to be happening in the town. The CCTV staff and police occasionally interacted to confirm that nothing was happening. Towards the end of the shift, however, a call came over the radio that seemed to be in code. It was from the police, it concerned a specific type of case and all eyes had to be on a local supermarket. The CCTV staff explained to me that this was a Mardi Gras alert.

A tip-off had been called in to the police and forwarded over the radio about a suspected Mardi Gras attack on a supermarket in the town. It seemed that in the event of a Mardi Gras attack, a protocol for viewing the town centre, collecting and compiling evidence had been set down in advance and this protocol was swiftly oriented towards by CCTV staff. At each step of activity, CCTV staff confirmed with police officers that the protocol was being followed. Various forms of interaction had been pre-planned; the staff were expected to perform tasks and report that those tasks had been completed. The real-time recorder was switched on and a full log was kept on the computer in the form of notes as to 'what was going on'.[6] The Mardi Gras bomber protocol involved following everyone in and out of the supermarket, the CCTV staff looked for bags, boxes or packages in the surrounding area and would have directed the police towards any claimed suspects or packages.

The protocol acted as the means to identify who should take responsibility for this threat (CCTV staff), the form of object which was to be singled out for attention (black rubbish bags), a transformation in the ontology of the object (from rubbish bag to potential bomb) and the ways in which activities should be reoriented to confirm this ontological transformation (CCTV staff should video particular things, pass these images on to the police and so on). This appears to be more straightforward as an ontological transformation than the somewhat messy example of the airport where passengers carried on with their usual activities despite ontological prompting. To some extent this is the case. However, the CCTV staff still oriented towards the protocol in ways that enabled them to continue with their usual activities. Although the CCTV staff were directed to look for suspicious individuals and packages, they spent a great deal of time focusing on a group of teenagers in the car park. Although such teenagers form one of their regular categories of suspicion (Neyland 2006), it was not clear that they fitted the

6 The CCTV system operated a time-lapse recording system the rest of the time which captured grainy images from every camera every few seconds and compiled these onto a single tape. The real-time recorder provided better quality and continuous images.

profile of lone bombers with a grudge. It was also curious that rather than any clear ending to the Mardi Gras alert, the CCTV staff gradually gave up looking at the supermarket and started to talk about and look at other areas of town. The extent to which the alert and the protocol transformed the ontological status of black rubbish bags and CCTV staff concern remained unclear. At most the protocol took up a few minutes of what was an otherwise dull Thursday shift.

The Mardi Gras bomber later delivered a final ultimatum in communication with police officers. The bomber demanded that money be left in a bank account and cards which could access the account with an agreed PIN number should be given away on the cover of *Computer Shopper* magazine. The police complied with this demand and put up to 1000 officers on duty ready to catch the bomber should any of the cards be used. The police later claimed that they would be alerted within three seconds of any attempt to withdraw money from the account. Police focused their operations on an area which the bomber had targeted during his campaign. Eventually the bomber was caught attempting to withdraw money from the account. Images of the bomber were captured on CCTV and by a media crew working with police officers. The Mardi Gras bomber was given 221 years in jail. At trial, no offence was ever mentioned in the town where the CCTV research took place.

ANALYSIS

This chapter has suggested that we need a greater understanding of the ways in which ordinary and everyday objects can be shifted into matters of concern. The approach adopted by this chapter has involved drawing together ideas from STS on the nature and status of objects and their networked relations, Foucauldian-inspired work on relations of governance and ethnomethodological insights into the contingent and constitutive nature of accountability relations. This combination suggests that we could think of objects' identities as being the upshot of the network of relationships in which they are entangled. That is, the shift of an object from an ordinary and everyday matter to a matter of concern is the result of the entities with which the object is engaged. Furthermore, it could be argued from this perspective that network relations are the focal point for the articulation of governance and accountability relations which seek to establish the new ontology for the object at the centre of network relations. In this way, new ontological identities for objects are suggested (for example, airport managers suggest bottles of water should be considered a terror threat rather than a drink), particular individuals and groups are identified as needing to take responsibility for managing this new threat (and this translates to, for example, passengers needing to be security ready), this responsibility involves recognising the need to reorient activities around the object (by not trying to carry it through security) and thus treat it differently, establishing its new ontological identity as a matter of concern.

Returning to the literature presented at the beginning of this chapter, we can see that ideas from ANT and more Foucauldian-oriented ideas of governance can be drawn together here in offering a description of what is going on. First, we can think of the object (say a letter) as the central feature of an assemblage (or network) of people and things through which the object's identity and identities of those members (human and non-human) which form the assemblage are constituted (or, at least, are somewhere in the process of becoming). In the ANT sense, this assemblage might be cut at certain points through organisational protocols such that, for example, airport security staff might disrupt networks to remove particular human (potential terrorists) and non-human (bottles of water) members of the network in order to retain the integrity of the assemblage. We could also think of these assemblages in more open or closed terms with letters, for example, potentially incorporating an assemblage comprising an entire organisation (all the members of the organisation who might be potential recipients of the letter) or black-boxed delegates of aggregate organisational representation (for example, through postal pigeonholes acting as proxy representatives for organisational departments).

This may seem a little passive. Although work is going on to accomplish an assemblage, we are not left with a strong sense of purposive effort. Second, through a more Foucauldian-inclined approach to governance, we might think of these assemblages as being held together and identities formed and reformed through what we could term relations of governance. We could then approach the directives from MI5 regarding potential letter bombs as the means to drive governance relations through the assemblage. Human and non-human members of the assemblage are designated specific roles to take on in recognising the potential new ontological status of the object at the centre of network relations and understand their identities in relation to the network as being conceived through the need to pay particular kinds of attention to the central object. Letters must be sniffed, closely inspected, perhaps x-rayed, not shaken too violently, opened only under certain conditions and so on. MI5 will not be in place to monitor this handling of letters. Instead it is the network which will form the focus of governance (for example, letter-handling roles and an understanding of the potential adverse consequences of letters for the organisation are delegated to particular features of the organisation such as post-room staff) and assessments of governance relations (with delegates assessed by other members of the organisation for the extent to which they have operated as adequate delegates of MI5 and the network's required governance. This might provoke questions in the organisation such as whether the post-room staff have done their job).

Although this provides us with general principles for thinking through the ways in which ordinary objects can become potential focal points of terror, it does not give us much detail on the ways in which this process might operate in practice. We can thus talk theoretically about assemblages establishing identities, governance and have some initial ideas on assessment, but does this adequately capture what is going on? It seems an incredibly tidy picture of events. The three examples drawn on in this chapter on counter security terrorist advice regarding letter bombs, objects of concern in airports and the role of CCTV in the story of the Mardi Gras bomber suggest that practices around objects are complex, messy and require their own specific attention. The example of objects in airports suggests that successive actions to build networks of governance around categories of objects (such as liquid containers and sharps), connecting various people (airport managers, passengers, security and check-in staff) and things (boards, plasma screen TVs, leaflets) in order to reorient actions around the object in focus and establish its new ontological status as a matter of concern are messy in practice. Airport passengers, in the view of airport managers, do not take time out from their journeys to look at boards displaying instructions on how to act towards new matters of concern, passengers are said not to look up at plasma TV screens and appear to, on occasion, ignore leaflets with instructions on what to take through security. The example of CCTV and the story of the Mardi Gras bomber suggest that even in relatively closed circuits of interaction between people (such as CCTV staff, and managers and police officers) and things (such as cameras, monitors, radios), attempts to reorient activities around new matters of concern (such as black rubbish bags) to affirm the ontological status of the object as potential threat, the extent to which that new status is accomplished remains mixed. The CCTV staff pay brief attention to the Mardi Gras protocol, begin to search for suspicious packages and people, then appear to lose interest and revert to their conventional categories of suspicion such as teenagers.

Returning to the ethnomethodological literature on accountability can broaden out our initial sense of assessment and help us in providing more flesh to the picture of what goes on in particular incidents of mundane terror. Although ANT and Foucauldian-inclined analyses can help us conceive of network relations and delegations of governance, we should not necessarily then assume that social actions such as governance relations are set once and then operate smoothly from then onwards. For every member of post-room staff who diligently does their job and acts as the perfect network delegate, others may forget, ignore or re-interpret various features of, for example, MI5 guidelines on potential letter bombs. In the alternative examples we find cases of airport passengers not paying much attention to their objects and CCTV staff not paying much attention to suspect packages. Governance relations do not fix subject positions but are a feature of on-going interactions which (in the ethnomethodological sense) continue to accomplish the activity in focus. Hence what constitutes adequate governance is played out through relatively messy on-going interactions which may be called into question only at moments which require further accountability scrutiny (for example if a letter bomb explodes, a passenger is called to one side by security or CCTV staff are questioned after an event for failing to spot something they were supposed to be looking for). Aside from those moments of further scrutiny which appear rare in the examples considered in this chapter, most relations of governance and accountability, drawn together through networks of relations articulated around particular objects, continue with a fluid sense of the object in focus (it is both its new and old ontology) and continue without the neat and straightforward relations some may seek (for example, airport or CCTV system managers).

Why does it appear to be the case that new routines to reorient understanding of the ontological nature of objects are so difficult to establish? It seems that ontologies are stubborn and routinised. Networks of relations between people and things have been well established around the ordinariness of the objects in focus (letters, scissors, water bottles, rubbish bags). Ordinariness fixes the object as being something which requires little attention; it is mundane, ordinary, even perhaps dull. Attempts to lift objects out of their ordinariness and into the category of things that now need attention require shifting stubborn and routinised network relations. The people and things connected around the object need to act and inter-relate in new ways. Attempts to introduce new governance and accountability relations (in the pervasive ethnomethodological sense), so that CCTV staff are held to account by police officers who are also following the protocol and airport passengers are held to account by signs, boards, screens, leaflets and airport security staff, are only partially transformative of the ontology of objects. The CCTV staff look at teenagers and lose interest in the protocol without much response from the police. The airport passengers carry on through to the security check carrying the wrong items, with only some passengers paying attention to signs and reorienting their actions towards objects accordingly. The airport security staff are retained as a vital filtering device. They operate to separate out people and their things or things and their people. This takes time and impacts upon airport profits. To reduce security queuing times and enhance profits, the role of airport security staff should involve merely acting as corroborators of separations that have already been performed by passengers.[7] However, this is only occasionally the case. Most of the security staff time and effort is dedicated to being more active separators of people and things. A consequence of the CCTV staff giving up looking for packages (and not being reprimanded by police officers) and airport passengers' not paying attention to new objects of concern (with security staff

7 This is in line with theories of governmentality and responsibilisation (mostly derived from Foucault's work; see Foucault 1977), i.e. that individuals should internalise rationales for the way they should act into their actions, thus displaying the extent to which they have taken responsibility for their actions.

performing the necessary separation) is that the old ontological status of the objects in focus and routines for actions which corroborate the original status, remain.

CONCLUSION

This chapter has argued that the surveillance-studies community can engage in a rich stream of surveillance activity by looking at the role of objects in everyday relations of governance, accountability and surveillance. The chapter has argued that objects can be conceived as focal points for networks of relations, held together by governance and accountability relations, which attempt to reorient actions around objects to confirm transformations of the objects' ontological status as no longer mundane but now a matter of concern. The chapter drew on three examples of counter terrorist security advice, objects moving through airports and CCTV protocols to look at the ways in which these objects and networks operate in practice. The chapter argued that attempts to shift the ontology of objects turned out to be complex in practice, as stubborn and routinised activities around the object and the original ontological status of the object retained their resonance. Further research in this area could usefully investigate other examples of objects of surveillance, provide further alternative models for understanding the surveillance relations between people and things, and the ways in which ontological shifts could be more effectively accomplished.

The research presented in this chapter suggests that security and insecurity can be played out through the mundane ontology of things. Routine notions of what things are, how they should be used and what their purpose is, builds and maintains a stubborn ordinariness for objects and their network relations. This ordinariness is at the heart of security matter(s): these taken-for-granted things need to be taken for granted in order to remain ordinary, unthreatening and ontologically secure. Terror threats which begin to undermine this secure ontology raise numerous concerns organisationally (for example in airports and CCTV systems) and in everyday interactions. However, unlike the horror film where we might have to kill our formerly beloved, recently zombified pet before it chews off our face, in these everyday incidents of mundane terror the ontological shift and necessary response are not so final or absolute. For every airport passenger who diligently buys into the terror threat posed by bottles of water and defers buying water until they have passed through security, there are others who continue to regard water as a drink and continue to guarantee long airport queues. Although one could argue that the queues themselves are a kind of mundane terror, it appears that in the case of objects, their ontological status as matter of concern is only ever one amongst several identities. Objects are thus simultaneously ontologically secure and insecure.

References

Adams, J. (2003) 'Seat belt laws: A clumsy perspective', paper presented at conference on 'Clumsy Solutions for a Complex World', University of Oxford, 5 April 2003.

Adey, P. (2004) 'Surveillance at the airport', *Environment and Planning A*, 36: 1365–1380.

Bash, H.H. (1995) *Social Problems and Social Movements: An exploration into the sociological construction of alternative realities*, New Jersey: Humanities Press International.

Baxter, J. and Chua, W.F. (2003) 'Alternative management accounting research: whence and whither', *Accounting, Organisation and Society*, 28(2): 97–126.

BBC (1999) 'Mardi Gra Bomber', available at: http://news.bbc.co.uk/1/hi/uk/317132.stm

BBC (1998) 'Why is the Mardi Gra Bomber so difficult to catch?' available at: http://news.bbc.co.uk/1/hi/uk/62434.stm

Bowker, G. and Star, S.L. (2000) *Sorting Things Out: Classification and its consequences*, Cambridge, MA: MIT Press.

Callon, M. (1986) 'Some elements of a sociology of translation: Domestication of the scallops and fishermen of St. Brieuc Bay', in J. Law (ed.) *Power, Action and Belief: A new sociology of knowledge?* London: Routledge and Kegan Paul, 196–233.

de Laet, M. and Mol, A. (2000) 'The Zimbabwe bush pump: Mechanics of a fluid technology', *Social Studies of Science*, 20(2): 225–253.

Drew, P. and Heritage, J. (eds) (1992) *Talk At Work: Interaction in institutional settings*, Cambridge: Cambridge University Press.

Ericson, R.V., Doyle, A. and Barry, D. (2003) *Insurance as Governance*, Toronto: University of Toronto Press.

Foucault, M. (1977) *Discipline and Punish*, London: Allen Lane.

Garfinkel, H. (1967) *Studies in Ethnomethodology*, Cambridge: Polity Press.

Grint, K. and Woolgar, S. (1997) *The Machine at Work: Technology, work and organisation*, Cambridge: Polity.

Gusfield, J.R. (1996) *Contested Meanings: The construction of alcohol problems*, Wisconsin: The University of Wisconsin Press.

Hetherington, K. (1997) 'Museum topology and the will to connect', *Journal of Material Culture*, 2(2): 199–218.

Hutchby, I. and Wooffitt, R. (1998) *Conversation Analysis*, Cambridge: Polity Press.

Latour, B. (1992) 'Where are the missing masses? The sociology of a few mundane artefacts', in W. Bijker and J. Law (eds) *Shaping Technology/Building Society: Studies in socio-technical change*, Cambridge, MA: MIT Press, 111–134.

Latour, B. (1991) 'Technology is society made durable', in J. Law (ed.) *A Sociology of Monsters: Essays on power, technology and domination*, London: Routledge, 103–131.

Law, J. (1994) *Organizing Modernity: Social ordering and social theory*, Oxford: Blackwell.

Law, J. (1991) 'Power, discretion and strategy', in J. Law (ed.) *A Sociology of Monsters: Essays on power, technology and domination*, London: Routledge, 165–191.

Lee, N. and Brown, S. (1994) 'Otherness and the actor-network: The undiscovered continent', *American Behavioural Scientist*, 37(6): 772–790.

Luff, P., Hindmarsh, J. and Heath, C. (eds) (2000) *Workplace Studies*, Cambridge: Cambridge University Press.

Miller, P. (1992) 'Accounting and objectivity: The invention of calculable selves and calculable spaces', *Annals of Scholarship*, 9(1/2): 61–86.

Miller, P. and O'Leary, T. (1994) 'Governing the calculable person', in A.G. Hopwood and P. Miller (eds) *Accounting as Social and Institutional Practice*, Cambridge: Cambridge University Press, 98–115.

Mol, A. (2002) *The Body Multiple: Ontology in medical practice*, London: Duke University Press.

Muller, C. and Boos, D. (2004) 'Zurich main railway station: A typology of public CCTV systems', *Surveillance and Society*, 2(2/3): 161–176.

Neyland, D. (2006) *Privacy, Surveillance and Public Trust*, London: Palgrave-Macmillan.

Neyland, D. and Woolgar, S. (2002) 'Accountability in action? The case of a database purchasing decision', *British Journal of Sociology*, 53(2): 259–274.

Norris, C. and Armstrong, G. (1999) *The Maximum Surveillance Society: The rise of CCTV*, Oxford: Berg.

Power, M. (1997) *The Audit Society*, Oxford: Oxford University Press.

Rappert, B. (2003) *Non-Lethal Weapons and Legitimizing Forces*, London: Frank Cass.

Rappert, B. (2001) 'The distribution and resolution of the ambiguities of technology or why Bobby can't spray', *Social Studies of Science*, 31(4): 557–592.

Rose, N. (1999) *Powers of Freedom*, Cambridge: Cambridge University Press.

Shove, E. (2003) *Things in the Making and Things in Action: A discussion of design, use and consumption* (SBS STS seminar, 24 January).

Rose, N. (1996) 'Governing "advanced" liberal democracies', in A. Barry, T. Osborne and N. Rose (eds) *Foucault and Political Reason*, London: UCL Press, 37–64.

Shove, E. and Southerton, D. (2000) 'Defrosting the freezer: From novelty to convenience', *Material Culture*, 5(3): 301–319.

Suchman, L. (1993) 'Technologies of accountability: Of lizards and aeroplanes', in G. Button (ed.) *Technology in Working Order: Studies of work, interaction and technology*, London: Routledge, 113–126.

Woolgar, S. (1996) 'Technologies as cultural artefacts', in W. Dutton (ed.) *Information and Communication Technologies: Visions and realities*, Oxford: Oxford University Press, 87–102.

Woolgar, S. and Pawluch, D. (1985) 'Ontological gerrymandering: The anatomy of social problems explanations', *Social Problems*, 32: 214–227.

Chapter 2

Identification practices

State formation, crime control, colonialism and war

David Lyon

INTRODUCTION

'Smart' national ID-cards systems are proliferating around the world, often sold as 'solutions' to security issues. Those adopting them believe that they offer certainty about the identities of people who wish to cross borders or make transactions and thus increase the national security of the societies in question. Critics and sceptics who question the workings of such systems, or even their very raison d'être, may ask about whether the certainty and security are as solid as the claims for them. It is not merely identities in the sense of individuating and distinguishing from all others, but also identities as members of particular groups and the registering of identities which are in question.[1] Such questioning, of both kinds, is part of my purpose here. Paradoxically, what some claim is certain and secure is from other viewpoints and for other persons both uncertain and insecure and produces uncertain and insecure experiences for those affected by them.

Some ID-card systems, such as those in Hong Kong and Malaysia, have been up and running for a few years, others, for example in Italy and Japan (Murakami Wood et al 2007), were more recently established and yet others are being trialled (India, China) or under development (UK, USA). Such systems have strong supporters, who argue for greater administrative efficiency and potential solutions to intractable problems such as illegal immigration, fraud, or terrorism. They also have detractors such as Genocide Watch, Privacy International and the Electronic Privacy Information Center, which warn about risks to privacy and civil liberties. A good way of considering these new technologies, associated with quests for greater security, is to see them as forms of surveillance. As with other kinds of surveillance, they focus attention on personal details for purposes of entitlement, access and policing (Lyon 2007: 13–16). And they are automated, draw data from the

1 It may be, too, that the very words do not translate very well: as Heidi Lomell pointed out to me, in German and Norwegian there is one word for both (*sicherheit*, *sikkerhet*, respectively (cf. Bauman 2000).

body, are local-and-everyday, and universal (Staples 2000: 4–7). They qualify as ideal candidates for Surveillance Studies.

Surveillance Studies is a relative newcomer on the academic scene and it is often rightly associated with the rapid spread of new technologies that depend on computing and communication power. But this intellectual child of the twenty-first century also harks back to earlier surveillance practices, some mediated by rather different information and communication technologies, which also demand to be understood. The new ID cards of today have a long and not always distinguished pedigree that should not be forgotten. Indeed, if we look at the case of large-scale identification systems for human populations, we can find examples of what Foucault called genealogies; both formative and failed schemes on the one hand and local knowledge and experiences on the other that may throw light on present-day developments (Dreyfus and Rabinow 1982: 119). Much may be learned from the history of the quest for stable IDs in the modern world that applies to the ongoing story in the present.

What links early and later efforts to provide stable identification systems is the demand for documents, to discover some reliable means of distinguishing the one from the many and of sorting the alien from the citizen or the imposter from the genuine. This is prompted partly by new travel possibilities that generate visitors not known through routine face-to-face contact. It was clearly the case that in early modern Europe the mobility of both messengers and migrant groups created problems for which identification documents were the official response, but the demand for ID documents went well beyond this. It is the official demand for documents of personal identity, for a range of related reasons, that links today's quest for new IDs with earlier practices. While claims may be made about the newness of ID-card systems, the continuities are actually quite striking.

Travel is one important factor. Modern modes of transport and communication permit travel of many kinds and for many purposes, but the proportion of populations that could travel and the processes of travel themselves have multiplied immensely since early modern days when transport and communication were still vitally linked. Communication depended on transport – the horse or ship that enabled the message to be transmitted – but the invention of the telegraph split these two apart. Communication could occur without transport. What is less remarked, however, is that there is a sense in which the relation has not disappeared so much as reversed. In a world of new technologies, transport now depends on communication. One cannot travel far without having to produce some marker or message that identifies and situates the traveller. Today, the information communicated often comes from the body of the traveller itself.

However, today's national ID-card systems have grown from a number of sites and sources, each of which contributes distinctive features to the processes of national identification and yet has a story of its own. Modern nation states rely on individuated forms of distinguishing one citizen from another and this involves both written registries and in some cases documents that must be carried on the person. But such states have also been involved in efforts to take particular note of certain classes of citizen and non-citizen, the undesirable, the deviant and the strategically useful. Thus identification systems have been used to discriminate between, curb the movement or block the entry of particular ethnic, religious or national groups; to specify which groups of lawbreakers or suspects are unwelcome or vulnerable to apprehension; to permit certain kinds of transaction or exchange; or to discover whose work or abilities might be usable or appropriated by the state for warfare or for production. Such features of ID systems represent continuities in the demand for documents.

In what follows, I trace some identification efforts aimed at making citizens more 'legible' (Scott 1998) within the 'embrace' (Torpey 2000) of the state and show how these have historically had to do with colonial administration, crime control and the exigencies of war. They have to do with both travel and transactions. In each case, the (in)securities and (un)certainties are writ large and have on several occasions had results that place bloody blights on human history. The certainties sought often involve clear categorisation – the imposition of classifications – that has facilitated sorting of actual populations, not merely for inclusion and exclusion but also for mass murder and genocide.

IDENTIFICATION AND THE LEGIBILITY OF CITIZENS

One process that distinguishes the modern nation state as such is the official attention accorded to individual details as part of the embrace of the individual by the state (Torpey 2000). John Torpey uses embrace in the sense of 'grasping' or 'registering' citizens in ways that both include and exclude particular persons (Torpey 2000: 12). In the telling trope used by James Scott (1998) such an embrace makes citizens more legible to the state and this in turn depended on both rising literacy and the growth of official records. However, it should also be noted that although the embrace of the state or the greater legibility of citizens may have deleterious effects, for example on levels of public trust, identifying citizens may also be the means of ensuring their entitlements and their rights.

Improving citizens' legibility may be undertaken for all kinds of reasons. In 1666 a detailed census was taken in Canada, largely for taxation purposes, but also to initiate an incentive scheme to encourage families to have more children (Ericson and Haggerty 1997: 111). Making citizens legible would increase not only revenue but reproduction, it seems. While John Torpey, in his work on the passport, shows how identification could help to create a monopoly on the means of movement (Torpey 1998), nation states also document identities in order to mobilise economic resources through taxation, to redistribute resources to citizens in need through welfare programmes and also through health and education, and finally to maintain peace and order.[2] The latter refers both to external threats from others powers or internal ones of rebellion, violence or crime. It is important to note that each initiative has a stake in adequate documentary identification and the parallel monitoring of populations. Identification has several interlinked purposes.

One of the most ancient reasons for registration and identification was to facilitate the taxing of citizens by the state and conversely to ensure that those eligible for state benefits received that to which they were entitled. Worlds touched by Christianity have no difficulty recalling that Jesus' birth coincided with a major tax-related registration under Roman rule. The Romans also used *tesserae* to identify slaves, soldiers and citizens. But in ancient China, 656–221 BC, tax and registration occurred in relation to war-making (Hui 2005). Similar systems existed in Greece and, even earlier, in Sumer.

Historians have little to say about identification processes in the ancient world, presumably because it was assumed in relatively fixed, local settings with little travel opportunity that by and large people were known to each other for most practical purposes. In early modern times things started to change, although even then such change was very slow and limited. As Edward Higgs (2004) has shown, identification systems appeared as part of a long-term, uneven process of rationalising state activities and these were generally expanded as needs arose and new techniques were produced that could make them more administratively efficient. Births, marriages and deaths, once locally registered in European or North American parishes, gradually became a state function. Such data would eventually become basic 'breeder' documents from which to create others.

Part of the problem was that in some cases surnames, the basic ingredient of any identification system, were not always stable. Many nineteenth-century immigrants to the USA and Canada, for instance, had no permanent surnames on arrival (Scott 1998: 71). Earlier cases of identification marks related not so much to immigration as to indigent people whose impoverished circumstances obliged them to move in search of sustenance. Though there were early cases of internal passports in sixteenth-century England, where the poor or vagabonds had to wear badges, this system did not last or develop (Higgs 2004: 42).

By the mid-twentieth century, citizenship – that involved registration but not necessarily a card carried on the person – had, as T.H. Marshall (1950) argued, expanded to include not only legal and political rights but economic and social ones as well. Documentary identification was required for each

2 This is a tongue-in-cheek reference to the principles on which Canadian confederation was built in 1867: 'Peace, Order and Good Government'.

kind of citizenship right to be maintained effectively. T.H. Marshall's studies of citizenship offer some important insights that still have a bearing on today's world, globalisation and neo-conservative restructuring notwithstanding (Isin and Turner 2007). Even if one queries T.H. Marshall's account of citizenship in its details, or criticises it for paying insufficient attention to the role of possessive individualism (MacPherson 1962), or acknowledges its need for updating for twenty-first-century situations, the ways that citizenship operated in Western Europe and North America were broadly beneficial. They incorporated populations within inclusive societies that choreographed a variety of rights and duties. At the same time, while such processes appeared congenial to majority populations, the growth of citizenship was not in all ways even and fair and indeed was contested periodically.

Gérard Noiriel writes about the 'révolution identificatoire' (Noiriel 1996) that produced various cards and codes for state identification purposes in the nineteenth century. These may have had a positive effect on those qualifying straightforwardly as citizens of France, but the use of such markers is also bound to cut both ways: the embrace of the state includes and excludes. Noiriel's analysis of the rise of the identity card in France and the treatment of foreigners based on certain methods of identification shows how social context and technology both play a role in producing a need for national identification. Such identification was at times used for discriminatory practices (Noiriel 1996: 60), which already rings warning bells about today's enhanced power – from biotechnology in particular – that could much more readily and profoundly produce discrimination, based on genetic and biological factors over which the individual has no control.

Such processes could be taken far further than they were in France. In the mid-twentieth century, infamous systems of internal passport were developed under the Nazi regime in Germany, in South Africa under apartheid, and in the Soviet Union. In Germany where, as Zygmunt Bauman (1991) has poignantly shown, the administration of the Holocaust represents the apogee of modernist rationality, International Business Machines (IBM) was recruited to provide the technical infrastructure for genocidal identification (Black 2001). In South Africa pass laws formed a kind of internal passport system that restricted the mobility and the life chances of black Africans. Interestingly, IBM, along with the UK firm ICL, also had a role in supplying the pass-book computer infrastructure from 1953. The hated books featured in protests and demonstrations against apartheid (although the ANC proposed another ID card system, HANIS, in 1996; see Breckenridge 2007).

Internal passports were used in Stalinist Russia and these too distinguished between desirable and undesirable populations. The Stalinist state sought to engineer society through either inclusion or exclusion, but the most detailed and comprehensive aspect of this identification, categorisation and monitoring was the internal passport and domicile registration system (Shearer 2004: 837–838). As well as documenting individual details, the passport was administratively linked to places of residence and work. It could thus be used to create a hierarchy of need for food and commodities in the times of scarcity that characterised the 1930s, and also for policing, state building and the larger state project of making socialism work. Categorisation by passport enabled the state to distinguish between 'threatening or alien (*chuzhie*) populations' and loyal ones, or ones 'close (*blizko*)' to the regime (Shearer 2004: 838). Shearer notes that the passport system could be read as a '. . . demographic and geographic map, literally, of Stalinist-style socialism' (Shearer 2004: 839).

The internal passport functioned in some ways analogously to the more familiar external passport system, only within the physical borders of the nation state. The passport gave the state an instrument for discriminating among its subjects in terms of rights and privileges. It regulated the movements of certain groups, restricted their entry into certain areas and denied them liberty to move away from their residential areas (Torpey 1998: 254). As Marc Garcelon points out, this amounted to a form of 'internal colonialism' entailing 'administrative differentiation' between citizens and subjects (Garcelon 2001: 84). The supposed dominant core has to incorporate a variety of culturally distinct groups if it is to survive, but those subaltern groups are likely to resist such incorporation.

Not only did the internal passport form the centrepiece of the Soviet surveillance and control system, its repercussions are also felt in the successor states that emerged after 1989 and the fall of communism. In 1995, for example, the mayor of Moscow was ordered to clear the city of unregistered persons from the Caucasus and Central Asia, using passport designations. Nearly 1 million Chechens – their passports stamped 'enemy of the people' – had been deported by Stalin in the 1940s and it was their children who fought for Chechnya in the mid-1990s (Garcelon 2001: 98). As Garcelon observes ironically, the internal passport maintained particular national identities rather than ever creating an internationalist and unified 'new Soviet man'.

Identification and colonial administration

Many early modern countries were involved in colonial rule, both internally with indigenous peoples or slaves and externally with overseas territories. In the American Old South, for instance, while slaves were officially denied their *identities* – they were treated as subhuman non-persons – their capacity for theft, arson or escape made them prime targets for *identification* (Parenti 2003: 14). The slave surveillance system on the plantations was based on what Christian Parenti calls three 'information technologies: the written slave pass, organised slave patrols, and wanted posters for runaways' (Parenti 2003: 15). In fact, in Virginia the earliest pass laws (1642) were directed against poor whites such as Irish indentured servants, but by 1687 South Carolina made such laws apply to black slaves. Literate slaves had an advantage (which is why slave owners usually tried to prevent slave access to education), as did those whose mothers had given them distinctive names so that they could keep track of their whereabouts. Black resistance heightened white resolve, however, and pass systems spread.

By 1793 South Carolina was using brass or tin tags – 'slave hire badges' – containing the name, date, occupation and number of the slave. The latter connected the slave to city records of payment of the annual slave tax (Parenti 2003: 25). Such cross-linkage with official records was extended with manumission, as the papers – still including written personal descriptions – distinguished between free and unfree subjects. The sorts of written descriptions available in the passes and papers were replicated in the wanted posters which as Parenti observes indicated both the nature of control and its limits; the slaves had already escaped. Some of the descriptive styles outlived the slave passes after the outlawing of the system, reappearing in early passports and in the control of immigrant labour.

A quite different situation of colonial administration existed in nineteenth-century India, where British rule included important experiments in identification. In 1858 Sir William Herschel, working with the East India Company in Bengal, initiated the first successful scientific forensic identification technique. As a member of the Indian Civil Service he wished to draw up a contract with a road construction materials supplier and asked Mr Konai to supply a handprint. He proposed that fingerprints be used more generally for legal documents but the idea did not fly. As magistrate of Hooghly, however, he instituted taking pensioners' fingerprints to obviate fraud, and those of prisoners so that they could not hire someone to substitute for them (Sengupta 2005).

This was not the end of the story, for Edward Henry and Azizul Haque of the Bengal Police perfected and systematised fingerprinting for forensic identification in 1893. While criminal groups and allegedly deviant populations were affected, the colonial state also kept detailed records of political intelligence on 'subversive activities' that extended to most domains of life. According to Sengupta, informers were cultivated in the 'criminal underground, the postal department, amongst railwaymen, soldiers, political activists, trade union members, lawyers, prostitutes, clerks, thieves, teachers, workers and students'. The Indian Telegraph Act (1885) permitted many kinds of state scrutiny of information.

Another very poignant example of colonial administration is that of the Belgian system of ethnic classification in Rwanda. It began with anthropometric measurements and ended with the issuance of obligatory identity papers stating one's ethnicity (Uvin 1997: 95). These papers became the means of genocide in 1994, based on fixed group identities, arranged hierarchically, and the fostering of distrust and hatred between the groups (Longman 2001: 346). Similar systems were developed throughout colonised countries of Africa. As in the American South, pass laws existed in the Cape Colony from the 1700s and Britain maintained this after their takeover in

1806. Mainly to regulate cheap and slave labour, such passes were common throughout British colonies in Africa and the practice was perpetuated within Belgian, French and Portuguese rule as well (although only in the former two cases did indirect rule using local indigenous leaders prevail (Longman 2001: 350)).

The original terms *Hutu, Tutsi* and *Twa* existed in pre-colonial Rwanda and Burundi and referred either to occupational or status distinctions that were flexible and could be changed. Europeans arriving in the late nineteenth century fitted these groups into new taxonomies that privileged Tutsis due to their supposed similarity (and thus superiority) to Europeans. Racial stereotypes, which assumed distinctiveness and mutual antagonism, were also developed for the other groups. However, when the Belgian authorities issued ID cards in the 1930s, it appears that the main aim was simply to regulate Belgian subjects and not to implement indirect rule through Tutsis (Longman 2001: 352). In practice, however, the newly fixed identities served to deny Hutus crucial opportunities for education and employment which in turn spawned the ethno-nationalist movements of the 1950s that denounced this subordination.

When the tables had turned in the 1990s, and Hutu military and political power was ascendant, the ID cards played a fatal role in determining who would live and who would die in the genocidal bloodbath. The compulsorily carried cards had to be shown at barricades and many Tutsis were killed on the spot when the documents were demanded. But some Hutus were also slaughtered because of their appearance, under suspicion of having false cards. Longman concludes, tellingly, that Rwandans had come to accept the principle behind the cards, '. . . that identities were fixed and unchanging, that everyone in the country could be clearly classified into one of three categories based on their parentage. It is this ethnicization of Rwandan society that ultimately made genocide possible' (Longman 2001: 356).

Identification and crime control

From earliest historical times means were found of marking lawbreakers. In the modern era, this was repeated in the New World. In the East Jersey codes of 1668 and 1675 burglars received a 'T' on the right hand and a forehead 'R' for a second offence, whereas adulterers in Puritan New England received the scarlet letter 'A' (Cole 2001: 7). By and large, markers were unnecessary in relatively immobile situations of local communities where most people were known to others. But with urbanisation and industrialisation came Simmel's 'society of strangers' (Simmel 1950) and this, along with the emerging machinery of municipal, state and national bureaucracies, offered the rationale and the means of marking populations singled out as lawless or troublesome. As nation states developed general systems of citizen registration, so specific kinds of identification were also sought for maintaining social order.

The quest for adequate means of criminal identification was sought constantly and ever more urgently during the nineteenth century, on both sides of the Atlantic. A good guide to this is the work of Simon Cole (2001). The reason was not simply the desire to identify correctly in relatively rare cases of imposture or mistaken identity but to find ways of coping with rising crime rates. It was widely believed that recidivism was a real problem, but without adequate identification it could not be proven that the same person had committed crimes repeatedly (Cole 2001: 13). Descriptions and photographs were circulated and in the UK an *Alphabetical Register of Habitual Criminals* was established. Increasingly, photos were sought as superior to written descriptions. With the emergence of the (no doubt hyped) 'confidence man' in the USA, the forger who could even deal in changed identities became the catalyst for better identification (Cole 2001: 21), especially that which focused on the face. Add to this physiological differences emphasised by the Italian school and the time-consuming quest for 'distinctive marks' used in Britain and one sees how the search for stable identities was considered urgent.

In the end, fingerprinting was to become the most widespread mode of criminal identification, despite great efforts to use Bertillonage, anthropometry and other modes of fixing identities. Problems of classification, to which Francis Galton made the chief contributions, were first worked out in colonial India and by the 1920s the superiority of fingerprinting over other modes was generally acknowledged. However, parallel developments in the USA and in Argentina – the former in contexts where the 'society of strangers' was more pronounced than in Europe – gave some fingerprinting a somewhat different cast. The connections between 'immigrants' and criminality were sometimes carefully traced in North America, for example with the influx of Chinese workers relating to the 1848 gold rush and transcontinental railway building boom (Cole 2001: 121). The 1882 Chinese Exclusion Act aimed at restricting the immigrant flow except in the case of workers who returned to China temporarily. But how to identify them? Once again, fingerprinting was soon preferred to physical descriptions and, as in India, the problem was viewed by officials as being compounded by the supposed difficulties of recognising one Chinese from another.

In Argentina, towards the end of the nineteenth century, Juan Vucetich, in charge of the statistical bureau of the La Plata police, developed further fingerprinting classification techniques that were also used in the categorising of immigrants. He produced his system of 'dactyloscopy' to simplify an earlier attempt to classify fingerprints on 101 variables. Vucetich added identification cards to the mix, thus indexing a large criminal identification file. But his work also furthered the cause of whitening Argentina, which had already succeeded in exterminating much of the Native American and severely reducing the African American population. What was once applied to black, whites and Indians was also applied to different European groups, some of whom were deemed preferable immigrants to others. Crime statistics had become confused with immigration issues and selective law enforcement strengthened the view that immigrants were primarily to blame for crime (Cole 2001: 132). These 'racial others' helped hasten the development and success of Vucetich's dactyloscopic scheme.

Interestingly, observes Simon Cole, early dactyloscopy was thought of less as a forensic than as a classificatory technique for linking bodies in custody to their criminal records. This also helps to explain the quest for a universal system that could eventually depend on remote records and that would, with the development of electronic storage systems, be used in conjunction with centralised identification bureaux. Moreover, it gives the backdrop to more than one proposal for universal fingerprint identification systems in the USA, for instance the 1943 Citizen Identification Act that called for ID cards with fingerprints (Cole 2001: 249). By the 1940s, however, IBM was supplying card sorters and punch-card systems for US fingerprint files. By the 1970s AFIS (Automated Fingerprint Identification Systems) were being used in the USA and elsewhere.

The practice of fingerprinting addressed a basic issue in all kinds of identification, not only criminal identification. While a surname is a prerequisite, and carrying an ID document may help the state authority to associate reliably a name with an individual person, that individual still must cooperate for the system to work. To achieve cooperation, states often make entitlement depend on the production of clear identification and in harsher regimes will punish individuals who fail to produce it. In situations of defiance, however, people will refuse or fail to identify or will dissemble and identify falsely. Fingerprinting has the advantage from a state perspective of being an ineradicable bodily mark, and the same argument would apply to most tattoos, other biometrics such as facial topography or iris scans and to DNA samples (see Scott 1998: 371 *n*38).

In the current drive for national ID cards various novel features appear such as radio frequency identification (RFID) tags, but almost all include at least one kind of biometric, fingerprints and digital photo images being the most popular. The argument is commonly used that while other forms of identification rely on what is known (and possibly forgotten), carried (and possibly stolen) or dependent on some code such as a PIN (that could be used fraudulently), biometrics relies on what we *are* (or at least what our bodies are at a given moment). Biometrics is usually credited with having superior economy, personal integrity, security, scalability and reliability. However, questions have been raised (for example in debates over the UK ID card) about potential racialised unevenness relating to Failure to Enrol (FTE) rates. Given the history of criminal and immigrant identification systems, checking in detail this aspect of new modes of identification and verification would seem to be a priority.

Identification for war

War has offered opportunities for creating identification systems in several modern societies. Both the need to discover who is fit, willing and available for military service and the need to distinguish effectively between loyal citizens and potentially hostile resident aliens have been behind mass identification programmes. More broadly, however, war and surveillance connected in other ways, that also help to contextualise specific dimensions of surveillance such as personal identification.

In general, it should be noted that the relationship between the military sector and surveillance has been underplayed. It deserves further historical and sociological investigation. As Christopher Dandeker has shown, the industrial and democratic revolutions in Europe and America catalysed the bureaucratisation of military power and therefore an expansion of their surveillance capacities (Dandeker 1990: 93). But the modern state not only controlled military power, the imperatives of modern war also helped to extend the bureaucratic surveillance that had developed in the military *back into* the wider society (Dandeker 1990: 101). The desiderata for a 'security state' spell increasing ties between the state and the military sphere, some of which work themselves out in the veil of secrecy regarding some specific operations and organisation and in the erosion of civil liberties relating to assembly, movement and the provision of information, areas in which the quest for new ID systems have been paramount (Dandeker 1990: 107; Lyon 2003).

Edward Higgs argues that the 'information state' in England was pushed forward significantly after 1914 by both military threats to Britain and the empire and the deepening and widening of the Welfare State (Higgs 2004: 133). Until this time the connection between military service and citizenship was not strong and no internal passports were used, as they often were in continental states. The population was brought into closer connection with the central state and personal data collection and analysis expanded accordingly. However, liberal traditions, modest aims and limited technologies meant that this expansion was not yet dramatic. The General Register Office (GRO) oversaw military recruitment data gathering and this was also linked with data on potential munitions, mining, railway and agricultural workers. A register of war refugees was also maintained to protect against fifth columnists and the GRO also sent information on enemy aliens to the intelligence services, MI5.

In Britain, the first national ID card appeared as the result of a fierce debate between supporters of conscription and those who wished to continue the voluntary method of obtaining recruits for the armed forces (Agar 2005). There had been a systematic official registering of births, deaths and stillbirths in Britain since the mid-nineteenth century, with a central repository of certificates held in London (information for this paragraph is mainly from Agar 2001: 104). Registers were also maintained of marriages, TB sufferers, voters, the mentally deficient, National Insurance contributors and primary and secondary school students. But no list was both universal and had up-to-date addresses. Beatrice Webb had campaigned for some time for a national ID card system, believing that progressive social reform would be served by such an innovation, but in the end it was the interests of 'industrial purposes' on the one hand and 'military and naval purposes' on the other that catalysed the National Registry into being.

By July 1915 the efforts under the National Registration Bill produced the results urgently sought by the War Cabinet, that almost 1.5 million men were still available for national service, but any broader aspirations associated with it remained unfulfilled (Agar 2005: 2). Not until September 1939 was the idea revived, now for 'national service, national security, and the administration of rationing' (Agar 2005: 3). A central National Register Office held records (near Southport) and ID cards were required for renewing ration books. They also were used for routine policing and this was the context in which they were eventually rejected for peacetime use. In 1950 a speeding motorist refused to show his ID card and his right was upheld in court on the grounds that the card was a war measure. The rationing registration system was transferred to the new National Health Service, but now without the ID card.

Continuity and change in large-scale identification

Whereas mobility offers part of the meaning to nascent ID systems and internal passports, it is not the whole story. When registration systems are connected with brands, badges or paper or plastic documents this certainly seems to facilitate the regulation of movement. However, the very act of identification and the growing requirement to produce proof has several profound social meanings, whether to indicate obligation or entitlement, to offer evidence of previous criminal records or none, or, even more importantly, to distinguish between those who are legitimate citizens, subjects or residents of a nation state and those who are not. The latter distinction is one that invariably rests on judgements about ethnicity and race, many of which are deeply prejudicial, divisive and exclusionary.

An aspect of the formation of modern nation states, especially by the nineteenth century, was the determination of what it meant to be French or British as distinct from other groupings. The 'essential nationhood' of one group was threatened by the possible invasion by non-national groups. In the British case, says Linda Colley, Catholics were suspect, as were peoples from colonies in India or Africa (Colley 1996: 37). Dangerous classes were associated through literary and imaginary ways with alien and uncivilised races, in contrast to good middle-class, empire-supporting identities. As Nikolas Rose puts it, following Foucault: 'The colonial experience and the codes of race were thus constitutively engaged in the formation of governable subjectivities . . .' (Rose 1999: 47). In this light, markers of identity and internal passports could be viewed as reinforcing the nation state as such by providing contrasts between legitimate and illegitimate identities. Workplaces and residential areas could thus be regulated, as could the territorial borders of the nation state.

Various kinds of identity documents appeared in the course of the twentieth century, from drivers' licences in the 1920s to social insurance cards and eventually health cards and credit cards (Rule 1974). While the *carte d'identité* existed in France or the *Personalausweis* in Germany, other documents – such as the driver's licence or the social security card in the USA – have been used as if they were ID cards. Even the Social Insurance Card, introduced in Canada in 1964, has served in this way, despite 'privacy' protests about function creep. The point is both that ordinary citizens have become increasingly accustomed to producing ID when documents are demanded, and that institutions have deepened their dependence on such cards, along with the corresponding registries and records.

It is important to observe that the categories used help to produce the citizens as such (Hacking 1990). Modern states, dependent on rational bureaucratic administration, are more or less bound by that fact to treat people according to their schemata. Schemes that started life as the inventions of census-takers, judges or police officers '. . . end by being the categories that organise people's daily experience precisely because they are embedded in state-created institutions that structure that experience' (Scott 1998: 83). Scott goes on to say that pass books, ID cards and the like '. . . acquire their force from the fact that these synoptic data are the points of departure for reality as the state officials apprehend and shape it' (Scott 1998: 83). If one needs a standing before the law, or to acquire entitlements, the classificatory documents provided by the state must be produced for the state.

During the later twentieth century, however, the logics of individualisation were carried much further and modes of social involvement multiplied. This was facilitated in part by the development of microelectronic technologies, searchable databases and networked computer systems. At the same time, and not unconnected with it, risk discourses were ascendant and communication of risk was vital for governance (Ericson and Haggerty 1997: 3). Identifying and managing risk is now a key task of contemporary institutions such as the police and insurance companies but also of government departments. Data gathered and communicated for this purpose includes, importantly, personal identification. But as Haggerty and Ericson say, the ('assemblage') system now works by '. . . abstracting human bodies from their territorial settings and separating them into a series of discrete flows. These flows are then reassembled into distinct "data-doubles" which can be scrutinized and targeted for intervention' (Haggerty and Ericson 2000: 606). From the point of view of the individual, self is fragmented.

The upshot of living in technology-dependent, risk-oriented and individualised social settings is that identification is required for more and more transactions, travel and communication. As Nikolas Rose notes, '. . . subjects are locked into circuits of control through the multiplication of sites where the exercise of freedom requires proof of legitimate identity' (Rose 1999: 240). Such identification 'inescapably links individuation and control' (ibid.). These processes of 'conditional access' to desired spaces (physical or virtual) Rose calls the 'securitizing of identity' within 'circuits of inclusion'. The converse, however, is the exclusionary processes within coexisting 'circuits of insecurity' (Rose 1999: 253). The universalising logic of the welfare state, called into question especially since the 1980s, has given way, suggests Rose, to an 'array of micro-sectors, micro-cultures of non-citizens, failed citizens, anti-citizens' that become the targets of risk management (Rose 1999: 261). But this was written before 9/11, when a further group – 'terrorists' – would be added to the list in ways that would help deepen the already existing divide in particular by racialising it further (see Lyon 2003). The post-9/11 world is one in which interest in identification and the demand for documents is intensified.

CONCLUSION

Today's quest for national ID card systems is not merely a reflex of technology companies seeking to sell 'solutions', or something specific to an era of post-9/11 panics and perceived problems of 'national security'. While it relates to both it is also the outcome of some long-term shifts that made plausible and facilitated the securitising of identity (and its correlate, insecurity and exclusion) and of some consistent themes, now reinterpreted in risk-ridden, global contexts, of ethnic and racialised others and remixed 'dangerous classes'. To recognise this contemporary search for national identifications as in some respects an old one is not to trivialise it so much as to acknowledge the genealogies that have helped to shape it as well as the political economies, cultural currents and geo-political forces that give it its peculiar present character.

While some identification documents such as passports exist primarily as markers of mobility (Torpey 2000), ID-card systems relate both to mobility and to entitlement and exchange (or 'transaction' (Isin and Turner 2007)). Thus in principle they combine several functions that have been found historically in discrete markers and registers, relating to military service and taxation on the one hand and social insurance, permanent residence and other benefits on the other. In so far as these may also be denied to some groups, however, it is worth noting the connections between excluded or suspect groups, especially where the boundaries are blurred between 'dangerous classes' and 'immigrants'. A key question, from the point of view of citizenship rights and responsibilities, is whether new ID-card systems will be able to escape the negative histories that have dogged such documents in the past.

Some other features of continuity and change in the quest of IDs that appear from the foregoing historical survey include these: a shift of emphasis from danger to risk, from order to security and from inclusion to exclusion. Schemes that once took root in the global south seem to shift as peoples from once-colonised countries migrate to the global north. The 'trusted traveller' has become an élite category for those who can afford fast-track passes. Identification in general has shifted from personal narratives and descriptions to body-data and dataveillance records. However, it is also important to move from these general descriptors to specific circumstances as things change and in some situations specific power pressures and negative experiences for particular groups proliferate. This is especially true of post-9/11 exclusion (where categories are generalised and blurred and it has become unclear whether the war on terror is in reality a war on immigrants (see Bigo 2005; Agamben 2005)).

While in the present context national ID cards are often sought (or sold) as 'security solutions', their other characteristics – above all their social-sorting capacities – are underplayed. Yet it is exactly such sorting that has been facilitated by ID systems far less technologically sophisticated and universal than today's. When the focus is on the card then virtues such as convenience and efficiency may be stressed. However, by focusing on the registry, the database and, by extension, the assemblage (Haggerty and Ericson 2000), social sorting, inclusion and exclusion come to the fore, along with their concomitant criteria that all too often relate to 'race' in ways that are more than reminiscent of previous ID-card systems.

The role of technologies companies and of softwares themselves should not be underplayed. What could be seen historically in the twentieth century with IBM in Nazi Germany and IBM plus the UK-based ICL in South Africa is now multiplied many times in the post-9/11 competition for security and surveillance contracts. IBM competed with other corporations such as Raytheon and Unisys to supply US-Visit, for example, losing to Accenture for the $7–10 billion contract. These companies state explicitly that post-9/11 opportunities are offering the chance for a revival of the flagging fortunes of ICT companies.

It is noteworthy that the older motifs of identification reappear in the new ID-card-system discourses. The colonial has given way to neo-colonial efforts to universalise identifications methods and to globalise some risk categories such as 'Middle Eastern' and 'Muslim' through 'interoperability' and the development of common standards. The crime-control motif for identification appears now in algorithmic methods, using data-mining in pre-emptive ways that include profiling. The new exigencies of 'war' in the global north have to do with an urgent but endless 'war on terror' and forms of 'national security' that are erroneously and egregiously associated with curtailing liberty. ID systems with their electronic registries and biometrics are easier to initiate in contexts of fear and secrecy. These three are also now interlinked in new ways, such that a criminalisation of immigrants is visible alongside a militarisation of crime control and the re-racialisation of 'enemy combatants'.

References

Agamben, G. (2005) *The State of Exception*, Chicago: University of Chicago Press.

Agar, J. (2005) 'Identity cards in Britain: Past experience and policy implications', *History and Policy*, paper 33. Available at www.historyandpolicy.org/archive/policy-paper-33.html/

Agar, J. (2001) 'Modern horrors: British identity and identity cards', in J. Caplan and J. Torpey (eds) *Documenting Individual Identity*, Princeton, NJ: Princeton University Press.

Bauman, Z. (2000) 'Social issues of law and order', *British Journal of Criminology*, 40, 205–221.

Bauman, Z. (1991) *Modernity and the Holocaust*, Cambridge: Polity Press (2001 Cornell University Press).

Bigo, D. (2005) 'Global insecurity: The field of the professionals of unease management and the ban-opticon', *Traces: A multilingual series of cultural theory*, 4: 34–87.

Black, E. (2001) *IBM and the Holocaust*, New York: Crown.

Breckenridge, K. (2007) 'Whatever happened to HANIS?' Paper presented at the research workshop on ID card systems at Queen's University, June.

Cole, S. (2001) *Suspect Identities: A history of fingerprinting and criminal identification*, Cambridge, MA and London: Harvard University Press.

Colley, L. (1996) *Britons: Forging the nation 1707–1837*, London: Vintage.

Dandeker, C. (1990) *Surveillance, Power and Modernity*, Cambridge: Polity Press.

Dreyfus, H.L. and Rabinow, P. (1982) *Michel Foucault: Beyond structualism and hermeneutics*, Chicago: The University of Chicago Press.

Ericson, R. and Haggerty, K. (1997) *Policing the Risk Society*, Toronto: University of Toronto Press.

Garcelon, M. (2001) 'Colonizing the subject: The genealogy and legacy of the Soviet internal passport', in J. Caplan and J. Torpey (eds) *Documenting Individual Identity*, Princeton, NJ: Princeton University Press. 83–100.

Hacking, I. (1990) *The Taming of Chance*, Cambridge: Cambridge University Press.

Haggerty, K. and Ericson, R. (2000) 'The surveillant assemblage', *British Journal of Sociology*, 51(4): 605–622.

Higgs, E. (2004) *The Information State in England: The central collection of information on citizens, 1500–2000*, London: Palgrave.

Hui, V. (2005) *War and State Formation in Ancient China and Early Modern Europe*, Cambridge and New York: Cambridge University Press.

Isin, E. and Turner, B. (2007) 'Investigating citizenship: An agenda for citizenship studies', *Citizenship Studies*, 11(1): 5–17.

Longman, T. (2001) 'Identity cards, ethnic self-perception and genocide in Rwanda', in J. Caplan and J. Torpey (eds) *Documenting Individual Identity*, Princeton, NJ: Princeton University Press.

Lyon, D. (2007) *Surveillance Studies: An overview*, Cambridge: Polity Press.

Lyon, D. (2003) *Surveillance after September 11*, Cambridge: Polity Press.

MacPherson, C.B. (1962) *The Political Theory of Possessive Individualism: Hobbes to Locke*, Oxford: Oxford University Press.

Marshall, T.H. (1950) *Citizenship and Social Class*, Cambridge: Cambridge University Press.

Martin, P. (2007) 'Inquiry kills off Access card', *Canberra Times*, 13 March. Available at: http://canberra.yourguide.com.au/detail.asp?class=lifestyle%20news&subclass=habitat&story_id=565277&category=finance/business

Murakami Wood, D., Lyon, D. and Kiyoshi, A. (2007) 'Surveillance in urban Japan: a critical introduction', *Urban Studies*, 44(3): 551–568.

Noiriel, G. (1996) *The French Melting Pot: Immigration, citizenship and national identity*, Minneapolis: University of Minnesota Press.

Parenti, C. (2003) *The Soft Cage: Surveillance in America from slavery to the war on terror*, New York: Basic Books.

Rose, N. (1999) *Powers of Freedom*, Cambridge: Cambridge University Press.

Rule, J. (1974) *Private Lives, Public Surveillance*, London: Allen Lane.

Scott, J. (1998) *Seeing Like a State: How certain schemes to improve the human condition have failed*, New Haven, CN: Yale University Press.

Sengupta, S. (2005) 'Signatures of the Apocalypse', *Metamute*. Available at: www.metamute.org/en/Signatures-of-the-Apocalypse/

Shearer, D. (2004) 'Elements near and alien: Passportization, policing and identity in the Stalinist state 1932–1952', *The Journal of Modern History*, 76: 835–881.

Simmel, G. (1950) 'The Stranger', in K.H. Wolff (ed.) *The Sociology of Georg Simmel*, Glencoe, IL: Free Press, lxiv, 445.

Staples, W. (2000) *Everyday Surveillance: Vigilance and visibility in postmodern life*, Lanham, MD: Rowman and Littlefield.

Torpey, J. (2000) *The Invention of the Passport: Surveillance, citizenship and the state*, New York and Cambridge: Cambridge University Press.

Torpey, J. (1998) 'Coming and going: On the state monopolization of the "legitimate means of movement" ', *Sociological Theory*, 16:3, 239–259.

Uvin, P. (1997) 'Prejudice, crisis and genocide in Rwanda', *African Studies Review*, 40(2): 91–115.

Part II

(In)secure spaces

Chapter 3

Spatial articulations of surveillance at the FIFA World Cup 2006™ in Germany

Francisco R. Klauser

INTRODUCTION

Between 9 June and 9 July 2006, the FIFA World Cup 2006™ (hereafter World Cup) dramatically changed public life in most German cities. In the media, the temporary reign of football over Germany's city centres has been most powerfully visualised through spectacular images of tens of thousands of mostly peaceful football fans on so-called 'public viewing sites' or 'fan miles', which was later named Germany's Word of the Year 2006.

Despite its concern with the football World Cup, however, this chapter is not about sport. Rather, this study focuses on another spectacular aspect of the World Cup, which can be exemplified by a simple number: 5.3 kilometres. In Berlin alone, fences 5.3 kilometres long and 2.2 metres high were erected, allowing the demarcation of an impressively large 'fan zone' in the city centre, reaching from the *Brandenburger Tor* to the *Strasse des 17 Juni*. Closely monitored by CCTV cameras, thousands of private security agents and police forces, this pre-defined fan zone – as the territorial framework for the concentration of fans on specific, and clearly separated, parts of the city centre – both materially and symbolically allowed the regulation of social life during the World Cup.

As a symptomatic illustration of the spatially bound logics of security and surveillance strategies, the picture of Berlin's fan mile also provides a powerful entry point to the main aim of this chapter which is to examine the territorial articulations of security/surveillance measures for the World Cup 2006. In this, the basic line of my argument is that security politics in general not only tends to relate to specific persons or social groups (Marx 1988; Lyon 2003) but also to select, classify, divide, mark, arrange, in one word, to differentiate specific categories of space. The functions of security and surveillance operations, their scope, impact and the risks they pose cannot be understood without referring to the territories concerned and created by their spatial deployment and performance.

Described as a cross-disciplinary, rapidly developing field of analysis and theory (Lyon 2002: 1), Surveillance Studies have sparked remarkable and revealing research over the last few years. Focusing on the increased possibilities of knowing, tracking, data-mining and profiling everyday life, one of the innovative powers of Surveillance Studies is to consider surveillance not only in relation to security issues but as a tool of governance in military conflict, health, commerce and entertainment (Haggerty and Ericson 2006). Recent work on surveillance thus provides a solid and fertile ground to examine the social implications of the proliferating range of new aims, agendas, objects, agents, technologies, practices and perceptions of surveillance from a wide range of perspectives.

Despite this increasingly sophisticated body of theoretical and empirical research, however, very few academics have provided critical accounts of the complex ways through which specific models of surveillance are becoming 'expert exemplars' for more normalised use. All too frequently, the study of particular surveillance projects is thus separated from the critical investigation of the broader processes, mechanisms and relationships, which lie behind the current proliferation of globally calibrated security procedures, operations and strategies.

It is from such a standpoint that this chapter engages with the spatial articulation of surveillance during the 2006 football World Cup, as a key moment, and as a key location, in the production and circulation of security/surveillance-related practices and expertise on different – local, regional or global – scales. The analysis builds upon the general understanding of the World Cup as both the product and the producer of a broader set of developments in security politics. Based on the study of a series of official documents (from police sources, political authorities and FIFA) and media articles about the World Cup, I shall advance a number of preliminary arguments in connection with four main developments at work within current dynamics and global re-calibrations of surveillance, which together constitute the basic structure of this chapter: the urbanisation, globalisation, technologisation and commercialisation of surveillance.

In each part, I shall first discuss how each development explains the relationships between the security/surveillance operations for the World Cup and space. I shall then provide a reading of how these developments are reflected in two specific examples of spatially anchored security measures: public viewing events and security rings around World Cup stadiums. From an analytic standpoint, this approach provides an exploratory framework not only to investigate where, by whom, how and to what purpose security politics imposed its logic on urban space but also to examine the broader processes at work within local, national and international interdependences in the co-production of security politics.

URBANISATION OF SURVEILLANCE

Mega sport events are typically moving from host city to host city (Hiller 2000). Their organisation and securitisation thus mainly constitute urban phenomena, even if their economic and social outputs are often expected to lie on a broader, national or international scale. According to Boyle and Haggerty,

> the primary fronts for security programs underwritten by recent developments are increasingly urban-centred. Security concerns are couched within, or coloured by, an urban frame of reference to the point that every security apprehension appears to be somehow urban and every urban issue is infused with security concerns. Mega-events figure prominently in the dynamics of this global re-calibration of security.
>
> (Boyle and Haggerty 2005: 4)

For the purpose of this chapter, emphasising the urban-centrism of mega sport events helps to explain both the general conditions and the specific needs as regards spatially anchored security/surveillance operations during the World Cup. In this perspective, however, the World Cup 2006 differs from other mega sport events in at least three important ways. First, in contrast to the concentration of Olympic athletes in specific villages near the host city of the Games, many national teams before and during the World Cup chose to stay in relatively remote villages, which often led to considerable security concerns in traditionally rural areas. In the small village of Achern, for example (in Germany's southern Black Forest region), English fans were allowed to camp near their national team's high-class residence, within a clearly designated area for up to 5000 fans (Dpa/Swr 12.4.2006: online). Rented out by a private provider, the camp was not only monitored by freshly installed CCTV cameras but also by both private and public security agents, aiming to secure the rearranged and demarcated (fenced) ex-parking field (Mühlfeit 20.6.2006: online). As we shall see shortly, these security measures are in many ways similar to security/surveillance operations for public viewing events in the urban environment. Second, the staging of the World Cup games affected not only one particular urban site but a network of 12 German host cities with World Cup stadiums, where most football fans, thousands of World Cup collaborators and hundreds of media representatives were concentrated: Gelsenkirchen, Dortmund, Cologne, Berlin, Munich, Hanover, Hamburg, Leipzig, Stuttgart, Frankfurt, Kaiserslautern and Nuremberg. Third, with more than 200 public viewing events in most large cities and many small villages, security issues became part of the agendas of various urbanised municipalities across Germany. Given the fundamentally different character and behaviour of football fans from supporters at Olympic Games, for example, and given the mobility of fans within the network of German host cities, security issues at the World Cup thus crucially affected large parts of Germany.

From this, however, we must by no means call into question the predominantly urban dimension of the World Cup's securitisation. On the contrary, considering security operations both in affected rural areas and in host cities as locally anchored key sites of the World Cup helps to identify the relationships between security politics and space more generally. In both cases, spatially anchored security operations were driven by the need to monitor and manage risks in a context of increased diversity and density, which obviously, but not only, applies to the urban environment, as the locus of increased density and diversity *par excellence*. 'Urban space gathers crowds, products in the markets, acts and symbols. It concentrates all these, and accumulates them' (Lefebvre 1991: 101). As I seek to demonstrate in this chapter, the spatial logics of security operations during the World Cup, both in cities and in rural areas, above all dealt with the marking, division, delimitation, i.e. with the differentiation, of relatively distinct and small portions of space, in order to regulate densely packed social activities through spatial operations and actions.

While this claim will be illustrated by the examples of public viewing events and security rings around World Cup stadiums, many other examples (from team hotels to railway stations, etc.) could in principle provide the basis for a more precise, micro-geographical analysis of physical and symbolic markings and arrangements of space by fences, patrolling police agents, access control installations, surveillance devices, etc.

Public viewing events and security rings around stadiums

The organisation of public viewing events constitutes a particularly meaningful example for pointing out how security politics, following the need to manage social risks in a context of increased diversity and density, translates into the urban territory. Conceived as central meeting spots for fans without match tickets, public viewing sites allowed supporters to watch football games on massive video screens in the heart of most German city centres. Clearly separated from their surroundings by fences, planned and often architecturally conceived like sport arenas (including different areas such as special children's sections and sections which were liable to pay costs), public viewing sites were in many ways treated like stadiums. Securing these publicly accessible 'places at risk' became one of the main focuses for both German and international police forces and for private security staff, which were hired by the commercial organisers of the events.

From this perspective, public viewing sites can be understood as the privileged spatial points of security politics within the urban environment, harbouring specific norms, values and constraints, including spot-checks of onlookers and specific legal regulations. According to these regulations, people with stadium bans or with a blood-alcohol level of more than 160mL were banned from public viewing sites (Gelsenkirchen 2006). In contrast to real stadiums however, access control to public viewing sites was not based on generalised identity checks.

Generally speaking, public viewing sites predominantly concentrated fans on specific points in the city centre. Thus, they hierarchically invested (selected, classified, separated, symbolically marked, materially arranged and controlled) particular portions of space, whilst other urban areas remained less considered. As we shall see later in this chapter, these differentiations of the city were further strengthened by the uneven deployment of surveillance technologies and by the reinforced presence of police and private security agents.

In addition to public viewing events, the so-called 'outer security ring' around World Cup stadiums provides a second, powerful example of how security politics resulted in new differentiations and hierarchisations of the urban environment, expressed as different types of constraints and stipulations. Reaching as far as 1 kilometre from the stadium (depending on the city), the outer security ring constituted the first clearly fenced barrier to the stadium for arriving fan groups. Restricted to holders of match tickets, accredited staff, members of the press and other authorised persons, the enclosed area was closed to the general public for the duration of the World Cup. The spatial delimitation of the outer security ring around World Cup stadiums, however, differs from public viewing events not only in its further restricted 'permeability' but also in its internal organisation. Conceived as the spatial stadium's extension, security rings were divided into four strictly separated sectors, following the need to avoid encounters between different fan groups. These sectors were accessible only by passing through particular access points, after repeated ticket and luggage checks. Despite these differences, however, public viewing events and security rings around stadiums are comparable in both their spatial logics and functions. Both cases bear material testimony of the production of distinct, spatial ensembles, which are materially and symbolically separated from their adjoining perimeters in order to regulate social activities through the separation and marking of hierarchically invested territories of security.

As the locus, medium and tool of security politics, and as an immediate, lived and experienced practical reality, both public viewing events and stadium rings can be seen as central points within the urban net- and meshwork of security politics. Serving to 'define both a scene (where something takes place) and an obscene area to which everything that cannot or may not happen on the scene is relegated' (Lefebvre 1991: 36), the securitisation of public viewing events and stadium rings above all relied on access control. Aiming to create safe and risk-free places by controlling flows (of people and objects), which are crossing the borderline between inside and outside at particular points in space, access control perfectly illustrates the fundamental spatial logic of security politics, which consists in selecting, classifying, differentiating, arranging and controlling specific portions of space, without according the same type of attention to the whole urban (or national) territory.

Above all, access control thus aims to guarantee the well-functioning of separated, differentiated and hierarchically organised parts of the urban environment, often carried to the point of complete segregation between indoor (secured) and outdoor (unsecured) space. Legitimised by the rhetoric of security, access control allows particular functions to be assigned to particular places. In other words, the regulation of social and spatial practices in the urban environment, at moments of increased social risks, does need borders and frontiers to control, organise, enlarge, facilitate, but also to supervise, enclose and if necessary repress. 'The prime function of surveillance in the contemporary era is border control. We do not care who is out there or what they are doing. We want to see only those who are entitled to enter' (Boyne 2000).

GLOBALISATION OF SURVEILLANCE

Surveillance, as the expression of a project, is the product of relationships, which are mediated by specific codes, techniques, intentions, domains of expertise, etc. It is thus of crucial importance to examine the networks of actors involved in the setting up, development and use of spatially anchored security strategies for the World Cup. At this point, however, it would be too tall an order to provide an exhaustive analysis of the whole panoply of actions and actors engaged in the securitisation of the World Cup. Rather, I will put particular emphasis on the proliferating range, scale and importance of multinational security collaborations.

To begin, it is worth providing some general examples of globalised security partnerships for the World Cup before examining the territorial expression of these linkages in relation to public viewing events and security rings around stadiums. Two years before the World Cup, Germany itself – together with Australia, France, Israel, Spain, the United Kingdom and the United States – took part in the Olympic Security Advisory Group, which provided coordinated security advice to Greece on its security planning (United States Government Accountability Office 2005: 6). During the World Cup, police officers from 13 countries were reported to join the German federal police, to build up the largest joint police operation in European history, as the spokesman for Germany's interior ministry was repeatedly quoted in the press (for example Associated Press 6.6.2006: online). Teaming up with German officers, international police agents were vested with similar competences as their German counterparts, including the power to arrest and expel fans of their own nationality. This collaboration was completed by intense exchanges of international hooligan databases, by close communications among secret services from different countries and by the integration of international terrorism experts (Bundesministerium des Innern 2004: 6). As the spokesman for Germany's interior ministry pointed out in the press, 'to give up that much sovereignty would have been unthinkable a decade ago' (Sachs 2006; cited in Associated Press 6.6.2006: online).

Furthermore, in order to control and restrict border crossing of 'undesirable' fans, bilateral agreements with all participating countries as well as with several neighbouring and transit states were signed before the World Cup. 'We want to create a threat filter which is effective beyond our borders: in the participating countries, transit countries and in the countries of our direct neighbours. The bilateral agreements are the basis for travel bans on hooligans and potential criminal offenders. They allow an intensive exchange of information and enable us to deploy security forces of partner states in Germany' (Schäuble 30.3.2006: online). Political agreements of such types are symptomatic of at least three broader developments in security politics: first, they again point out the crucial importance of access and border control (and thus of mobility management) for security politics, in addition to the already mentioned range of inner-urban access controls. As I have argued above, border control relates to space in a most significant way, in that it constitutes and reinforces a spatially anchored system of limits, resulting in the arrangement, marking and differentiation of space into hierarchically organised territories of security. Second, Germany's bilateral agreements with participating and neighbouring countries point towards the high relevance of formalised, transnational alliances within security politics. If we want to uncover the relationships embedded in contemporary developments of security politics, and if we want to assess how security systems are subsequently planned, built up and used, the importance of government alliances on a global scale cannot be underestimated. Security issues at mega sport events are thus often described as a catalyst in setting off much broader and longer-lasting international security collaborations (Chan 2002). Third, the temporary and flexible reintroduction of border controls with Schengen partners strongly underlines the current exemplification processes of security politics. In recent years, this measure has indeed become a common 'exemplar' of dealing with security and terror issues at major (sport) events. For example, Portugal re-introduced border checks during the European Football Championships 2004, while Finland did the same during the 2005 World Athletics Championships in Helsinki.

Fourth, the emphasis on joint international security operations for the World Cup can be highlighted by Germany's request for the assistance of two Nato Airborne Warning and Control System planes (Awacs), in order to provide airspace surveillance for this 'Special Major Event', as the World Cup was called in military jargon (Bittner and Klenk 11.5.2006: 10). This request not only gives another flavour of the scale and importance of the securitisation of the World Cup on an international scale but also emphasises the broader trend towards the increasing militarisation of public safety, linked to the prevention of crowd violence and terrorist attacks (Warren 2004). There is another point to be made here regarding the aforementioned exemplification of security politics. Since Nato began to give air surveillance support in 2001, as part of the Alliance's contribution to the defence against global terrorism, Awac planes have flown more than 3000 hours for more than 30 events, including the Summer Olympic Games in Athens 2004, the 2005 Winter Games in Turin and the Pope's visit to Poland a few days before the World Cup (Nato 6.6.2006: online).

Public viewing events and security rings around stadiums

More particularly, in order to understand the transposition of current globalisation processes of security politics onto the level of urban morphology, it is worth looking back at the examples of public viewing events and security rings around World Cup stadiums. Public viewing sites indeed constitute a powerful example of the increasing globalisation of security politics, i.e. its co-production between numerous public and private, local, national and international parties. While local (commercial) organisers were held responsible for the securitisation of public viewing events in the first place (Polizei Nordrheinwestfahlen 2006: 3), public viewing sites also constituted the privileged territorial framework for the deployment of the above mentioned national and international police forces, as various images on internet weblogs (for example www.flickr.com) of posing fans with English, French or even Angolan police officers suggest. Globalisation processes of surveillance can thus not only be seen in connection with the planning and setting up of security politics 'behind the scenes' but also within the urban environment itself.

The same applies to security rings around World Cup stadiums, revealing again to what degree security politics during the World Cup has been co-produced through globalised, public–private security partnerships, adding to the high number of state actors (from police agents to fire brigades and emergency services) more than 15,000 private, nationally and internationally recruited security agents and stewards were employed by FIFA for an estimated €30 million, mainly for security purposes within the outer security rings of the stadiums and for ticket controls (Borchers 17.5.2006: online).

Adding to the erection of fences, as new material and symbolic borderlines in the urban environment, the deployment of international security personnel strongly contributed to the demarcation and control of specific spatial ensembles, as the privileged locus, medium and tool of security politics. Both public viewing events and stadium security rings acquired normative value as an immediate practical reality only through their active surveillance and regulation by globalised and privatised security partnerships and by the wide use of surveillance technologies, as we shall see in the following section. The degree of differentiation which was superimposed upon the urban environment above all followed the mobilisation of myriad different actors, harbouring specific domains of expertise, instruments, etc. Only if we take into account the various needs and intentions of these parties (both behind the scenes and on the spot) can we understand how spatially anchored security measures are helping to impose a certain order on the urban environment.

Both public viewing events and stadium security rings also indicate how the security measures employed and tested during the World Cup can be setting new trends for security politics more generally. In the 'Host City Charta' for the organisation of the 2008 European Championship in Austria and Switzerland, a detailed contract between UEFA and the European Championship host cities, the staging of public viewing sites and the demarcation of stadium security rings is indeed prescribed with great care and explicit reference to the FIFA World Cup. Swiss and Austrian police delegates closely followed every step of their German homologues during the World Cup (Blick 26.5.2006: online). Public viewing events and stadium security rings can thus accurately be described as pre-defined security models, which are based on the delimitation, demarcation, material arrangement and symbolical marking of particular portions of space within the urban environment.

TECHNOLOGISATION OF SURVEILLANCE

Security issues at mega sport events also involve the increasingly complex assemblages of disconnected, semi-coordinated and heterogeneous forms and functions of surveillance (Haggerty and Ericson 2000). Mega sport events are indeed largely used as test sites of increasingly sophisticated high-tech security, thus strongly pushing forward the use of new, preventive arrangements of control and surveillance, which are disproportionately valuing the surveillance and securitisation of particular 'places at risk'. This claim is powerfully exemplified by experiences in Athens' summer Olympics 2004 (Samatas 2006), Turin's Winter Games 2005 and Germany's FIFA World Cup 2006. In Athens, the so-called 'C4I-system' included thousands of computers, surveillance cameras (partially equipped with automated behaviour-recognition software) and microphones (able to analyse dozens of languages). This unprecedented science-fiction security system was modelled on a range of military technologies including underwater sensors, patriot missiles, zeppelins and US battleships. During the World Cup, the 'nerve centre' for German-wide security operations was located inside the Interior Ministry in Berlin. Here, 120 security agents, equipped with monitoring screens, dozens of computers and sophisticated communication gear, brought together satellite views, close-up CCTV images from sport arenas and city centres and reports from police sources, the military and from intelligence services (Nickerson 7.6.2006: online). On-the-spot, specialised police agents employed 'fast identification' devices for DNA analyses of suspect individuals (Bild 11.6.2006: online).

Importantly, these examples not only underline the aforementioned globalisation of security and surveillance issues but also point towards the multiplication of private responsibilities in providing technologically based solutions in matters of public safety and counterterrorism policies. The growth of socio-technical arrangements and operations that are put to work within security politics is also resulting in new interdependences between different parties concerned. Here I have in mind in particular the technical competences required to manage high-tech security systems, which are likely to give certain highly specialised, private parties more weight. The multiplication of the use of socio-technological mediations in the 'making' of surveillance is also leading to new procedures and even leading to new, highly specialised professions, such as 'surveillance designers' for example (Ruegg, November and Klauser 2004). Furthermore, it is particularly interesting to note that the use of surveillance technologies is accompanied not only by the creation of an increasing number of private intermediaries but also by the development of a specialised language, the use of which becomes accessible only to specialists. Relationships between the user- and the supply-side of surveillance technologies are thus going far beyond the level of mere business relations in that they are bringing together a wide range of subtle, complex and contingent interests, strategies and reciprocal implications.

Public viewing events and security rings around stadiums

The growing use of technologically based security operations at mega sport events probably finds its clearest expression in the securitisation of the outer security ring around World Cup stadiums. Digital communication technologies in the rebuilt Olympia Stadium in Berlin included nearly 300 kilometres of cabling, converging in the stadium's Facilities Management Centre as the heart of the security system. Here, private security staff and police agents jointly monitored CCTV images of the stadium, the underground car park and the routes to the boxes. From this central point of the security system, most of the monitored locations in Berlin could also be visualised, as well as transmitted images from mobile surveillance vehicles within the city centre.

Furthermore, a large number of other high-tech security devices were used by security staff, such as robots to check the stadium's surroundings for bombs before matches and high-resolution cameras with face-recognition software, allowing the recording of biometric facial features of suspected hooligans which could be checked in real time against photos stored in a central database (Blau 29.5.2006). Importantly, all 3.5 million match tickets were sold with embedded RFID chips, containing personal information on the ticket holder (name, address, date of birth, nationality and number of ID card or passport), which was electronically checked not less than four times before arrival at the stadium.

Besides these spatially bound and technically based access control measures, it is particularly relevant to focus on CCTV, in order to point out the spatial logics of surveillance in general and for the World Cup more particularly. We have already seen that the organisation of the World Cup resulted in the first use of biometric face-recognition cameras in Germany. Consider as well the condition of CCTV monitoring for the staging of official public viewing events in the whole of Germany, leading to the implantation of an important number of additional surveillance cameras in public places in German host cities. In Stuttgart, for example, hundreds of CCTV cameras were installed for the World Cup, provided by the same manufacturer (IndigoVision) and with similar technical features, as for the 2006 Olympic Games in Turin and in Athens in 2004. Adding to this, public transport companies in many German cities just before the World Cup invested millions of euros in CCTV technology, such as in Munich, where an additional 542 surveillance cameras were installed two years before the World Cup to monitor metro stations, escalators, etc. (Münchner Verkehrsgesellschaft 4.2.2004: online).

In fact, while much effort has been expended on analysing video surveillance as a tool of social sorting, there is a current lack of research regarding the spatial logics and characteristics of CCTV. Before targeting specific social groups or individuals, the installation points of the cameras, their technical features (zoom, angle of vision, etc.), their direction while unattended and the active manipulations of their position by camera operators are first and foremost related to specific portions of space. Individuals or social groups are monitored once they enter the cameras' gaze. Social behaviour is of interest only within the cameras' premises. As a limited window to the city, video surveillance must thus above all be considered as 'surveillance of space'. First, the camera's position can be quite vertical in order to concentrate on one particular point in space, often corresponding to access gates or entrance doors. Second, the monitoring of certain 'spatial points' may be enlarged to 'spatial lines'. In this case, the cameras' gaze not only allows coverage of one particular point of interest but the monitoring of whole building walls, platforms in metro stations, etc. Consequently, the camera's position will be more horizontal, following the need to 'stretch' its field of vision. Third, in the case of movable, swivelling and zooming cameras, CCTV might enable the transmission and recording of visual information, relating to larger 'spatial surfaces'. Once again, however, surveillance operations will be restricted to specific parts of space. Corresponding to different spatial scales of surveillance, all three types of CCTV powerfully illustrate the logics of security politics to select and to disproportionably monitor distinct, hierarchically organised and relatively small portions of space, with the result of new spatial disparities between more or less monitored areas within city centres.

COMMERCIALISATION OF SURVEILLANCE

A growing body of theoretical and empirical research focuses on the value of mega sport events as 'entrepreneurialist' strategies of public policy (Harvey 1989; Hubbard and Hall 1998), entitled to promote cities' and nations' tourist image (Hannigan 1998; Fainstein and Judd 1999), to facilitate urban regeneration, to attract financial investments and, consequently, to produce economic development (Euchner 1999; Degen 2004). As we see in official statements from the German government, the same logic also applied to the World Cup, which was presented by the Interior Ministry as a unique opportunity for a 'business location and image campaign' to promote Germany as both a 'hospitable, cosmopolitan and modern country' and a 'strong and innovative place' (Schäuble 30.3.2006: online).

Yet this highly revealing literature on mega sport events in terms of city marketing and 'place selling' (Philo and Kearns 1993; Horne and Manzenreiter 2006) tends to ignore completely the business-relevant role of security politics. Consider, by way of example, Konrad Freiburg, head of the German police union, who stated that 'there would be terrible pictures seen all over the world – in which 200 mad neo-Nazis are being protected by a ring of 1000 policemen from a counter-demonstration. This would be shameful. It's not the image of Germany we want to present' (Freiburg 2006; cited in Furlong 5.6.2006: online). In this light, threats of terrorism and escalating hooligan or neo-Nazi violence were seen not only to endanger the population but also to threaten the carefully constructed marketing image of an 'enjoyable, colourful and secure World Cup' (Schäuble 30.3.2006: online).

However, the World Cup's economic appeal cannot be reduced to its importance as a business location and image campaign for Germany and its host cities. On the contrary, the World Cup above all constituted the commercial product of a powerful, profit-oriented global player: the Fédération Internationale de Football Association (FIFA). Even if the World Cup was financially supported by the German government, the 'Länder' and the host cities, and even if the event was hosted by the German Football Association (DFB), it was officially organised by FIFA. 'This is not Germany's World Cup, but FIFA's World Cup' – FIFA president Joseph Blatter was famously quoted in the press (Hanimann, 16.6.2006: online). This statement is of major importance for the last part of this analysis, as it also raises significant issues regarding the relationships between the tremendous security efforts during the event and FIFA's business interests in the World Cup, wherein I will concentrate on FIFA's subtly forwarded attempts to guarantee the exclusive branding of city space by its official sponsors. In this, public viewing events and stadium security rings provide a particularly meaningful illustration of the complex relationships between security politics, economic policy and private business interests, or – in other words – of the relationships between processes of securitisation and branding of space.

Public viewing events

During the World Cup, the spatial delimitation and differentiation of public viewing sites corresponded not only to functional differences and to different security standards but also to different degrees of commercialisation between the inside and the outside. On the one hand, fences around public viewing events separated and marked specific 'places at risk', which became the object of increased control, based on security technologies and realised through globalised, public–private security partnerships. On the other hand, the same fences also marked the spatial limits of FIFA's sphere of influence within the city, given the fact that FIFA fully controlled the organisation and marketing of public viewing events. Indeed, these fan festivals helped to push forward FIFA's power to produce its own, commercially useful urban environment in at least three ways.

First, public viewing events principally had to be registered and licensed by the Swiss company 'Infront Sports', FIFA's television partner and the holder of all public viewing rights in Germany (Martens et al 9.3.2006: online). Furthermore, depending on the classifications of the event as commercial or non-commercial, public viewing licences were liable to pay costs. Second, FIFA fully controlled the symbolical marking of fan festivals through the prescription of brands and advertisement boards to be displayed. In this, prominence was given to the logos and products of FIFA sponsors (Wilson 6.6.2006: online). Only in non-host cities were other sponsors admitted, as long as they would not be competitors to official FIFA partners. In this way, FIFA succeeded in creating a 'clean', commercially useful environment for its official partners' products and advertisement banners. Consequently, many of the most prominent urban squares in German city centres were invested by FIFA interests for the duration of the World Cup. The public viewing site in Cologne, for example, on the famous Roncalliplatz, offered splendid views not only of the Cathedral but also of the prominently positioned Hyundai exposition model beside a large screen. Third, FIFA also managed the spatial dimensions of public viewing events, their separation from the surrounding urban environment and their internal subdivisions and arrangements. For instance, although public viewing events strictly concerned public space, FIFA had to give its approval for any extension of the events' size, to comply, for example, with the wish of many cities after the first round of the World Cup (Stadionwelt and dpa 16.6.2006: online).

Security rings around stadiums

Security rings around stadiums provide a second example to consider the spatial concurrences between security and business interests. Before the World Cup, the whole outer security ring had to be handed over to FIFA as 'neutralised space', with all signs of advertising and sponsorship removed. The early, legally binding agreement, determined in point 8.1 of FIFA's specifications for the organisation of the World Cup in the so-called 'FIFA Pflichtenheft', had to be signed by the German government and by each host city before it was even known whether Germany could organise the event (Pfeil 3.11.2005: online). In this agreement we can deduce not only the weight of sponsors' needs within the organisation of the World Cup but also the deep connection between security issues and appeals of mass marketing. Spaces near the stadiums had to be separated from their surroundings not only to provide risk-free games but also to provide the privileged stage for branding and advertisement strategies and thus to become commercially invested (symbolically marked and materially arranged) by FIFA sponsors.

At this point, the question might arise as to whether there is any way of dating what may be called the origin or driving force behind the spatial concurrence between security and business interests. A definitive answer to this question would obviously require more detailed and comparative empirical investigations into the complex processes, mechanisms and relationships in the setting up and staging of mega sport events. However, if indeed there is a need at all to identify a 'first step' and to maintain its distinction from the 'following steps', its importance must be quite relative. Related to the Actor Network Theory as developed by Bruno Latour and Michel Callon (Latour 1992), there is good reason to assess the 'making of stadium security rings and public viewing events' – or the 'making of Mega Sport Events' more generally – as the result of complex, subtle and highly interwoven interactions and interdependences of myriad different actors, strategies and interests. Various questions and interests are in play in the set-up of surveillance-based security politics, just as many different aspects help to model the ability of particular measures or constructions to respond to the existing demands.

At this stage, it is of most importance to underline the factual correspondences between different functions of space, which allowed FIFA to impose its own spatial rationality and commercial branding within the re-territorialised stadiums' surroundings. Both our examples of spatially bound security operations during the World Cup – public viewing events and security rings around stadiums – thus point towards the fact that the partitioning of the urban environment into specific areas of control also stood for specific relationships to the city, mediated through FIFA's intentions to create a clean environment for its official partners' merchandise.

Following on from this, it is particularly interesting to note that the negative implications of stadium security rings for concerned residents and local business companies have been widely annihilated, discursively, by the supposed usefulness of stadium security rings (as their very name suggests) for security purposes, thus fading out potential critiques of FIFA's economic benefits. For example, in order not to compete with FIFA sponsors' interests, local car garages had to remove their advertisement marks (because of the exclusivity of Hyundai as official FIFA sponsor) and restaurants had to hide their outside beer signs (advertisement reserved to Budweiser). Furthermore, to guarantee the FIFA sponsors' exclusivity, seven of 12 stadiums were renamed 'FIFA World Cup stadiums' because their original denomination contained the name of a commercial company. In Munich and Hamburg, the huge sponsors' names outside the stadiums even had to be removed by a crane (Wilson 6.6.2006: online). Yet FIFA did not only control the materiality of the stadiums' surroundings and names but also fans within the arenas. Before the game between the Netherlands and Ivory Coast, for example, FIFA collaborators found Dutch fans guilty of ambush marketing because of the logo of a Dutch beer company – which was not one of the official FIFA sponsors – on their orange dungarees. Consequently, hundreds of fans had to take off their trousers before entering the 'security ring' around the stadium.

CONCLUSIONS

This chapter was driven by two broad objectives. On the one hand, it was concerned to critically examine the FIFA World Cup 2006 in Germany as both the product and the producer of a general cluster of developments in security politics: the urbanisation, globalisation, technologisation and commercialisation of security/surveillance issues. Yet these developments do not enter into antagonism with each other. On the contrary, each development embodies and nourishes the others. For example, the predominantly urban-centred proliferation of high-tech surveillance technologies also highlights current trends in security politics which is becoming increasingly global in scope (addressing globalised social risks and bringing together globalised security partnerships) and commercial in nature.

On the other hand, and through the lens of these four developments, the chapter has analysed the relationships between security politics and space and the production of hierarchically organised 'territories of security' within Germany's city network during the World Cup more particularly. These investigations not only repeatedly underlined the logics of security politics to select and classify specific portions of space, to separate these places from their surroundings and to symbolically mark, materially arrange and control these portions of space, but also highlighted FIFA's attempts to reconfigure the urban environment into relatively small, disproportionally commercialised spatial entities.

In order to assess the issues that are linked with this enquiry, we have to remember how deeply space is related with society. On the one hand, space is produced by society and its inherent relationships of power. In this regard, the last part of the chapter pointed towards FIFA's power to 'hegemonically' produce its own commercially useful urban environment during the World Cup. On the other hand, space produces society. According to Lefebvre, a decisive part is played by space in the continuous reproduction of society. 'Space commands bodies, prescribing or proscribing gestures, routes and distances to be covered. It is produced with this purpose in mind; this is its raison d'être . . . Space lays down the law because it implies a certain order – and hence also a certain disorder (just as what may be seen defines what is obscene)' (Lefebvre 1991: 143). We must, in this light, understand the final aim of the interwoven processes of selection, classification, separation, symbolical marking, material arrangement and control of spatial entities as the regulation and control of social activities. Pointing towards the spatial logics of security politics thus also highlights the linkages between security politics, space and social relationships of power more generally.

Exemplifications of security politics

The emerging picture of this twofold analysis suggests a series of further investigations into the roles and wider social implications of mega sport events as an important research programme within the interdisciplinary field of Surveillance Studies (Lyon 2007).

First, there is a crucial need to further investigate the increasing importance of private actors and commercial goals within current developments of security-driven, spatial reorganisations of the urban environment. While the linkages between the increasing commercialisation of urban space and the proliferation of spatially anchored security measures such as CCTV, for example, have been subjected to repeated analytical scrutiny (Reeve 1998; Coleman and Sim, 2000; Töpfer, Hempel and Cameron 2003), very few academics have provided critical accounts of how these developments are pushed forward by globally operating business companies (the FIFA and its official World Cup sponsors, for example). In this, it seems particularly worthwhile to further investigate the pressures to differentiate city space into clearly demarcated 'territories of security and commerce', arising from sponsors' interests, private insurance companies, but also from locally anchored shops, hotels, etc. Or, to put it as a question: what type of commercially motivated interests, practices and relationships lies behind the security operations and strategies which assume to protect the population from dangers?

Second, my study of the securitisation of the World Cup points towards a series of important 'issues of scale', which might guide future empirical investigations of the interactions and interdependences between global, regional and local security partnerships. What can mega events tell us about the interactions between security issues on different – local, regional and global – scales? How do global security partnerships relate to and intervene in particular local circumstances?

Third, there is a strong and pressing need for further empirical investigations into the contribution of mega sport events – as test sites for the use of complex high-tech surveillance systems – for the development of increasingly standardised ways of dealing with security issues more generally. In both scholarly research and public debate about current developments in security politics, there is in fact almost a complete silence on the question of how specific security measures are becoming expert 'exemplars' for more normalised use not only in similar circumstances ('horizontal exemplification') but also in other, more trivial moments, situations and places of everyday social life ('vertical exemplification'). In the first case, mega sport events must be further exploited in their importance as a privileged locus, where globally operating standard actors – moving from country to country, from city to city and from event to event – are implanting increasingly standardised security solutions to create standardised territories of security. The underlying assumption could be that the potential applications of these standard security solutions are not defined in relation to any locally anchored social, cultural or legal specificities but by the predefined equation: specific type of event = specific range of possible applications of security models. In this regard, critical attention must above all be paid to the increasingly important part which is played by private security companies, 'wandering the planet in search of consultancy fees and places to save, "parachuting in" to localities with plans and designs and then moving on to the next place – almost as if they float free without any connection to any kind of territory' (Holden and Iveson 2003: 66).

In the second case, referring to the 'vertical exemplification of security politics', this standardisation process does not only apply to structurally similar places, moments and events. Rather, previously tested security solutions (such as RFID chips or biometric face-recognition software for access control to sport stadiums, for example) also tend to be generalised in more ordinary places, situations and moments of everyday life (such as in supermarkets, etc.). From this perspective, it will be of major importance to further evaluate the new international pressures arising from internationally pre-established security models, which are increasingly influencing local decisions. What does this development – which could also be described as an increasing 'normalisation of the exceptional' (Agamben 2005; Flyghed 2002) – mean in terms of the scope for critical democratic debate about the appropriateness and proportionality of specific surveillance measures?

Regarding the FIFA World Cup 2006 in Germany more particularly, a major issue will be whether the temporarily engaged security measures will continue to impose themselves within the urban environment. What future will be reserved for those fences, checkpoints and technological infrastructures whose installation was legitimised by the exceptional circumstances of the World Cup? On a political level, and after the emotions evoked by the event itself, these questions should be resolved calmly, by considering again the wide range of social costs and benefits related to the above shown trends in security politics.

References

Agamben, G. (2005) *State of Exception*, Chicago: University of Chicago Press.

Associated Press (6.6.2006) 'World Cup Security Force Assembles in Germany', *Fox News*. Available at: http://www.foxnews.com/printer_friendly_story/0,3566,198410,00.html (accessed 15 October 2007).

Bild (11.6.2006) 'Gen-Tests für deutsche Hooligans', *Bild*. Available at: http://www.bild.t-online.de/BTO/news/aktuell/2006/06/11/gen-testhooligans/gentest (accessed 15 September 2006).

Bittner, J. and Klenk, F. (11.5.2006) 'Die Mannschaft für die schlimmsten Fälle', *Die Zeit*, 11 May, 20: 10–11.

Blau, J. (29.5.2006) 'World Cup – security scores big at tournament', *Computerworld*. Available at: http://www.computerworld.com.au/index.php/id;1926576695;relcomp;1 (accessed 15 October 2007).

Blick (26.5.2006) 'WM als EM-Testlauf', *Blick*. Available at: http:www.blick.ch/sport/wm06/artikel37580?layout=popup (accessed 15 September 2006).

Borchers, D. (17.5.2006) 'Fussball-WM: Zwickmühle Sicherheit', *heise*. Available at: http://www.heise.de/newsticker/meldung/print/73232 (accessed 15 October 2007).

Boyle, P. and Haggerty, K.D. (2005) 'Spectacular security: Mega-events and the security complex', paper presented at the *Our North America: From Turtle Island to the Security and Prosperity Partnership* Speaker Series, hosted by the Department of Political Science, University of Alberta, Edmonton.

Boyne, R. (2000) 'Post-panopticism', *Economy and Society*, 29(2): 285–307.

Bundesministerium des Innern (2004) *Die Welt zu Gast bei Freunden*, Dritter Forschungsbericht des Stabes WM 2006 zur Vorbereitung auf die FIFA-Fussball-Weltmeisterschaft 2006, Berlin: Bundesministerium des Innern.

Chan, G. (2002) 'From the "Olympic Formula" to the Beijing Games: Towards greater integration across the Taiwan Strait?', *Cambridge Review of International Affairs*, 15(1): 141–148.

Coleman, R. and Sim, J. (2000) 'You'll never walk alone: CCTV surveillance, order and neo-liberal rule in Liverpool city centre', *British Journal of Sociology*, 51(4): 623–639.

Degen, M. (2004) 'Barcelona's Games: The Olympics, urban design, and global tourism', in M. Sheller and J. Urry (eds) *Tourism Mobilities: Places to play, places in play*, London: Routledge: 131–142.

Dpa/Swr (12.4.2006) 'Die Briten kommen doch nach Baden', *Sport ARD*. Available at: http://sport.ard.de/wm2006/wm/vorort/swr/news04/england_camp.jhtml (accessed 15 October 2007).

Euchner, C.C. (1999) 'Tourism and sports: The serious competition for play', in D.R. Judd and S.S. Fainstein (eds) *The Tourist City*, London: Yale University Press, 215–232.

Fainstein, S.S. and Judd, D.R. (1999) 'Global forces, local strategies, and urban tourism', in D.R. Judd and S.S. Fainstein (eds) *The Tourist City*, London: Yale University Press, 1–20.

Flyghed, J. (2002) 'Normalising the exceptional: The case of political violence', *Policing and Society*, 13(1): 23–41.

Furlong, R. (5.6.2006) 'Hosts tackle security risks head-on', *BBC News*. Available at: http://news.bbc.co.uk/1/hi/world/5008296.stm (accessed 15 October 2007).

Gelsenkirchen Glückaufkampfbahn (2006) *Fanfest-Ordnung*. Available at: http://www.glueckaufkampfbahn2006.de/down/Fanfest-Ordnung.pdf (accessed 15 October 2007).

Haggerty, K. and Ericson, R. (eds) (2006) *The New Politics of Surveillance and Visibility*, Toronto: University of Toronto Press.

Haggerty, K and Erkicson, R. (2000) 'The surveillant assemblage', *British Journal of Sociology*, 51(4): 605–621.

Hanimann, D. (16.6.2006) 'Deutsche lassen Blatter links liegen', *Blick*. Available at: http://www.blick.ch/wm06&artikel39054?layout=popup (accessed 15 September 2006).

Hannigan, J. (1998) *Fantasy City*, London: Routledge.

Harvey, D. (1989) 'From managerialism to entrepreneurialism: The transformation in urban governance in late capitalism', *Geografiska Annaler*, 71B: 3–17.

Hiller, H. (2000) 'Toward an urban sociology of mega-events', *Research in Urban Sociology*, 5: 191–205.

Holden, A. and Iveson, K. (2003) 'Designs on the urban: New Labour's urban renaissance and the spaces of citizenship', *City*, 7(1): 57–72.

Horne, J. and Manzenreiter, W. (eds) (2006) *Sports Mega-Events: Social scientific analyses of a global phenomenon*, Oxford: Blackwell Publishers.

Hubbard, P. and Hall, P. (1998) 'The entrepreneurial city and the "new urban politics" ', in P. Hall and P. Hubbard (eds) *The Entrepreneurial City*, Chichester: John Wiley & Sons, 1–26.

Latour, B. (1992) *Aramis ou l'amour des techniques*, Paris: La Découverte.

Lefebvre, H. (1991) *The Production of Space*, Oxford: Blackwell.

Lyon, D. (2007) *Surveillance Studies: An overview*, Cambridge: Polity Press.

Lyon, D. (ed.) (2003) *Surveillance as Social Sorting*, London: Routledge.

Lyon, D. (2002) 'Surveillance Studies: Understanding visibility, mobility and the phonetic fix', *Surveillance & Society*, 1(1): 1–7.

Martens, R., Gertz, H. and Greulich, M. (9.3.2006) 'Widerstand der Fan-Guerilleros', *Spiegel online*. Available at: http://www.spiegel.de/sport/fussball/0,1518,druck-404181,00.html (accessed 15 October 2007).

Marx, G.T. (1988) *Undercover: Police surveillance in America*, Berkeley/Los Angeles: University of California Press.

Mühlfeit, P. (20.6.2006) 'Fancamp zur "Chefsache" erklärt', *Sport ARD*. Available at: http://sport.ard.de/wm2006/wm/vorort/swr/news06/20/englandcamp.jhtml (accessed 15 October 2007).

Münchner Verkehrsgesellschaft (4.2.2004) *Die neue U-Bahnbetriebszentrale der MVG: High-Tech-Steuerung für 300 Mio. Kunden*, Press report. Available at: http://www.mvg-mobil.de/presse/presse_2004/04.02.2004b.htm (accessed 15 October 2007).

Nato (6.6.2006) 'NATO support for FIFA World cup in Germany'. Available at: http://www.nato.int/shape/news/2006/06/060606a.htm (accessed 15 October 2007).

Nickerson, C. (7.6.2006) 'Security measures kick into high gear', *The Boston Globe*. Available at: http://www.boston.com/sports/articles/2006/06/07/security_measures_kick_into_high_gear?mode=PF (accessed 15 October 2007).

Pfeil, M. (3.11.2005) 'Platz da für die Fifa!', *Die Zeit*, 45. Available at: http://www.zeit.de/2005/45/FIFA-Republik (accessed 15 October 2007).

Philo, C. and Kearns, G. (1993) 'Culture, history, capital: A critical introduction to the selling of places', in G. Kearns and C. Philo (eds) *Selling Places: The city as cultural capital, past and present*, Oxford: Pergamon Press, 1–32.

Polizei Nordrheinwestfahlen (2006) *Sicherheit bei der WM – grundsätzliche Zuständigkeiten*. Available at: http://www1.polizei-nrw.de (accessed 15 September 2006).

Reeve, A. (1998) 'The panopticisation of shopping: CCTV and leisure consumption', in C. Norris, J. Morran and G. Armstrong (eds), *Surveillance, CCTV and Social Control*, Aldershot: Ashgate, 69–88.

Ruegg, J., November, V. and Klauser, F. (2004) 'CCTV, risk management and regulation mechanisms in publicly-used places: A discussion based on Swiss examples', *Surveillance and Society*, 2(2/3): 415–429.

Samatas, M. (2006) 'Security and surveillance in the Athens 2004 Olympics: Some lessons from a troubled story', paper presented at the International Sociological Association World Congress, ad hoc session on Security, Surveillance and Social Sorting, July.

Schäuble, W. (30.03.2006) 'We are creating the basis of an enjoyable, colourful and secure World Cup', presented at Conclusive Security Conference on the 2006 FIFA World Cup in Berlin, 30 March. Available at: http://www.bmi.bund.de/cln_012/nn_772756/internet/Content/Nachrichten/Reden/2006/03/BM__WM__Sicherheitstagung__en.html (accessed 15 October 2007).

Stadionwelt and dpa (16.6.2006) 'FIFA: Längere Fanmeilen erlaubt – Pläne in mehreren Städten', *Stadionwelt*. Available at: http://www.stadionwelt.de/wmspecial/index.php?template=news_detail&news_id=263&stadion=Allgemein (accessed 15 October 2007).

Töpfer, E., Hempel, L. and Cameron, H. (2003) *Watching the Bear: Networks and islands of visual surveillance in Berlin*, Working Paper no 8, Urbaneye RTD-Project, 5th Framework Programme of the European Commission, Technical University Berlin: Centre for Technology and Society.

United States Government Accountability Office (2005) *Olympic Security: U.S. support to Athens Games provides lessons for future Olympics*, Report to Congressional Requesters, no GAO-05'547. Available at: http://www.gao.gov/new.items/d05547.pdf (accessed 15 October 2007).

Warren, R. (2004) 'City streets: The war zones of globalisation: Democracy and military operations on urban terrain in the early twenty-first century', in S. Graham (ed.) *2004, Cities, War and Terrorism*, Oxford: Blackwell, 214–230.

Wilson, B. (6.6.2006) 'Stadiums renamed for FIFA sponsors', *BBC news*. Available at: http://news.bbc.co.uk/1/hi/business/4773843.stm (accessed 15 October 2007).

Chapter 4

Checkpoint security

Gateways, airports and the architecture of security

Richard Jones[1]

INTRODUCTION

The aim of this chapter is to suggest a general theoretical model of 'checkpoint security'. My central argument is that checkpoint security is a specific kind of control practice within crime control and criminal justice, finding various applications in police stations, at security roadblocks, prisons, courts and national borders, but also more widely in society, at airports, underground railways, ports, schools, mail rooms, galleries, offices, military facilities, shops, gated communities, and even pubs and clubs – indeed anywhere where it is thought important to regulate those passing through. In many respects, security checkpoints are simply a form of situational crime prevention (used to 'increase the effort', by controlling access to facilities or by screening exits), but I will argue that their usage is sufficiently widespread to be deserving of criminological attention in their own right. One could argue too that security checkpoints are merely a particular form of surveillance practice. In fact, security checkpoints can be seen to bridge situational crime prevention and surveillance practices, suggesting a new way of conceptually linking these two areas together. I aim to identify features shared by security checkpoints with the aim of building up a general sociological model of their operation.

I will argue that checkpoint security typically exploits constraint measures in order to process people passing through the checkpoints and that 'architecture' is at the heart of checkpoints. The constraints exploited are typically of one of two kinds: physical limits of the body, which are exploited in terms of *channelling* individuals along channels and through gateways; and *checking*, performing checks on people, their bodies, identities and belongings.

1 Acknowledgement: I would very much like to thank the editors, and in particular Katja Franko Aas, for their very helpful and perceptive comments on earlier drafts of this chapter.

SECURITY CHECKPOINTS: THE SOCIAL AND POLITICAL CONTEXT OF THE POLICING OF BOUNDARIES

If a gateway is an opening in a bordered place that may be opened or closed, a checkpoint is a way of regulating the flow through a gateway, and a security checkpoint is a form of checkpoint focused on security. Not all checkpoints are focused on security, ticketing checkpoints being one such example. We can think of checkpoints as ranging across a spectrum, from the minimal security conferred by ticket authentication to maximum-security checkpoints. Security checkpoints represent a very particular type of response to perceived social and political threats of 'unsecured' flows of people across borders (see Franko Aas 2005). As such, their use takes place against the backdrop of flows of people across spaces. These can include the spaces of regions, nations, cities, institutions, facilities or shops. The use of security checkpoints generally can be situated within a wider social geographic context, including increased geographic mobility, economic internationalisation, and advances in information and communication technologies. Regulation of access to places can be used to try to control access to valuable information or privileged places, to consolidate existing social advantage or to confer it anew. Information technologies may lead to the reconfiguration of earlier dependencies on places, in favour of 'the space of flows' (Castells 1989), but also heighten the lure of specific places as key nodes for economic reasons or for the lifestyle they promise (see for example Castells 2000; Florida 2004; 2005; Hayward 2004). Whatever the actual extent of these flows, and whatever their true motivation, they would appear to include illicit flows, both of people and of things (Franko Aas 2007a; 2007b; van Schendel and Abraham 2005; Nordstrom 2007). Linking places of attraction are 'non-places' of transit (see Augé 1995), but punctuating the grey worlds of these non-places, and inspecting their traffic, can be found the nodes of security checkpoints. Often, checkpoints are used in such ways as to regulate entry into a place, but sometimes (for example, in shops, libraries, warehouses, prisons or offices) they are used, also or instead, to regulate exits.[2] A distinctive feature of gateway securitisation is a focus (for the most part) on those attempting to pass through, whereas threatening individuals 'at large' within the general population are mostly left unaddressed. It is only when individuals present themselves at a gateway that they come under particular scrutiny.

Security checkpoints serve to allow (some) through; they regulate the flow from one side to the other; as Schneier notes, 'all security systems need to allow people in, even as they keep people out' (2006: 181). In their selective exclusion, they could be seen as a central device in perpetuating wider social exclusion. Jock Young has written about what he sees as the exclusionary

2 I am most grateful to Heidi Mork Lomell for this point.

tendencies of late modernity's consumer capitalism (1999; 2007), but as he also recognises, 'Physical, social and moral boundaries are constantly crossed in late modernity' (2007: 31). While 'a characteristic of late modern society is the setting up of barriers, of exclusion', this does not bring about a simple social isolation, since 'the virtual communities set up by the mass media easily transcend physical demarcations':

> The binary language of social exclusion fundamentally misunderstands the nature of late modernity. Here is a world where borders blur, where cultures cross over, hybridise and merge, where cultural globalisation breaks down, where virtual communities lose their strict moorings to space and locality. The late modern city is one of blurred boundaries . . .
>
> (Young 2007: 31)

This is not a process of exclusion alone: 'Rather it is one where both inclusion and exclusion occur concurrently – a *bulimic* society where massive cultural inclusion is accompanied by systemic structural exclusion' (2007: 32).

Rather than merely to exclude, checkpoints may be established in the name of 'security'. The drive for security takes place against the backdrop of a perceived world of insecurity, a world in which significant flows of people make some fearful of introducing dangers 'from the outside', or of allowing dangerous networks to develop. If 'surveillance' is the attempt to know about and hence gain power and advantage over a population 'out there', with the aim of governing it more effectively and efficiently (Foucault 1979), the use of regulated gateways marks a different kind of regulatory tactic – that of waiting for a segment of the population to present themselves before the checkpoint. Foucault discusses the 'discipline-blockade' as a precursor to panopticism (1979: 209), but we can perhaps regard gateways more as a endeavour developing in parallel to surveillance forms, rather than preceding them. In lectures bridging his studies on 'discipline' and 'governmentality', Foucault draws a distinction 'between security and discipline': whereas discipline seeks to normalise ('one started from a norm'), the 'apparatuses of security' pragmatically attempt instead to deal with what is out there ('starts from the normal') and introduce the 'new notions' of '[c]ase risk, danger, and crisis' in the attempt to manage the emerging 'problem of circulation' (Foucault 2007: 55; 63; 29; 61).

Checkpoint security and the 'surveillance society': simply surveillance?

There is now a well-established and impressive body of research on the role of surveillance in contemporary societies and on what has been termed 'surveillance society'. Taking, for instance, the particular example of airport security,

Lyon has argued that the implementation of airport security practices post-9/11 should be seen less as a distinctly 'new' and more 'as part of a long-term trend and against the background of the emergence of a surveillance society and a safety state' (2006: 398). These are useful notions, can be linked to Garland's (2001) 'culture of control' thesis, and help characterise and conceptualise societal responses to perceived social threats, intrusions and disorders. Lyon notes that recent airport security innovations in Canada include the introduction of the 'Advanced Passenger Information/Passenger Name Record program (API/PNR)':

> Under the provisions of the Immigration and Refugee Protection Act that came into force in 2002, commercial carriers are required to provide Citizenship and Immigration Canada (CIC) with passenger and crew information for analysis, so that any who appear to pose concerns may be identified and intercepted. Such data include five elements: full legal name, gender, date of birth, nationality, and travel document number. In a novel move, airlines must provide CIC with such passenger and crew data before they arrive in Canada. Under joint Canada-U.S. agreements, various means such as CANPASS Air (the Canada-U.S. harmonized, iris-scan-dependent, NEXUS Air scheme) have been implemented (November 2004) to expedite the travel of 'approved, low-risk travellers'.
>
> (Lyon 2006: 400)

Lyon argues that this sort of development is part of a wider trend in which the

> key practice here is that of producing coded categories through which persons and groups of persons may be sorted [Cayhan 2005; Lyon 2003b]. If personal data can be extracted, combined, and extrapolated in order to create profiles of potential consumers for targeted marketing purposes, then, by a similar logic, such data can be similarly processed in order to identify and isolate groups and persons that may be thought of as potential perpetrators of 'terrorist' acts.
>
> (Lyon 2006: 404)

The kinds of developments to which Lyon points are no doubt happening in many countries and may well, as he argues, be part of a wider social process. Other writers such as O'Malley (2006) have discussed the use of 'risk' assessments in airport passenger screening, and Adey (2004a; 2004b; 2006) has written a series of interesting pieces about airport surveillance and 'the relationship between mobility and practices of surveillance' (2004b: 1365).

However, the argument of this chapter is that whilst all this may be true, in addition to the surveillance of air-travelling individuals through registration, biometrics and database solutions, airport security and surveillance typically involve situated practices of security and surveillance, exploiting physical qualities of bodies and baggage. Moreover, while, as Adey notes, 'airports are symbols of mobility' (2004: 500), when seen at close hand they sometimes appear somewhat slower places, including the slow-moving queues found ahead of each of a series of checkpoints. Many of the surveillance practices involve scanning slow-moving or even stationary objects or people, and of then channelling these through certain pre-defined and limited channels. Many such practices bear as much similarity to situational crime prevention as to surveillance. Whilst security checkpoints present an opportune location at which surveillance can be overlaid (see also Hobbs et al 2003: 121, *passim*; Lyon 2007: Chs 2, 6, *passim*) and seem to exhibit conceptual parallels as well as real linkages with what Lyon and others have identified as 'surveillance as social sorting' (Lyon 2003b), they should not be reduced to surveillance alone. Instead, security checkpoints can be seen as carrying out other, situated, social-controlling roles within a wider social geographic context of managing flows of people, or of responding to perceived insecurities (Lyon 2007). In security checkpoints, what are being socially sorted are not just data but also real bodies and things.

Security concepts and their role in checkpoint security

It is useful at this point to introduce a few security concepts. In the context of a discussion about the use of biometric security technologies, Prabhakar et al (2003) distinguish between *verification* and *identification*. In the former mode, what is being checked is that the person's details or biometric measurements match those previously obtained from or allocated to them. In this case, the details offered as verification are checked against the one database entry for that person's name, to verify that the details match, to establish validity. In the second mode, however, the identification details supplied are checked against the whole database, seeking a unique match, to establish identity. Schneier (2006: 182–183; *passim*) suggests a further category, distinguishing between identification (establishing a person's identity), authentication (the same as verification, above, establishing proof of who they claim to be) and authorisation (establishing what the person is allowed to do). As Schneier notes, in the real world, 'many systems jumble identification, authentication, and authorization' (2006: 182). Conceptually, however, the distinction is a very useful one for thinking about what security systems are or should be trying to achieve. Checkpoint security may be used to police one or more of these modes.

IMPLEMENTING SECURITY: HUMAN, BUREAUCRATIC AND TECHNOLOGICAL SOLUTIONS

As Schneier (2006) and Dror (2006) note, the oldest way of implementing security is by human sight – establishing a person's identity by recognising them from their face, or by recognising an item such as a seal, ticket or document as genuine. As societies developed over the centuries, states introduced various bureaucratic means to try to assist institutions and checkpoints in establishing the identity of individuals, cataloguing these identities in archives, and issuing individuals with official documentation (Torpey 2000). In relation to 'papers' generally, Torpey proposes a threefold typology, differentiating between international passports, internal passports (as were used in South Africa during apartheid, and in the former Soviet Union) and identity cards (which may, among other things, 'be used by the authorities to enforce intermittent checks on movement') (2000: 165). His distinction draws attention to the different roles the different forms of papers play. An international passport, for example, is sometimes used as a means of identification, but can also provide authorisation (for example, to allow the bearer to enter a particular country without further checks). Identification papers, however, can be subverted; a person may obtain another's papers and impersonate them; papers may be entirely faked; or otherwise genuine papers may be doctored with a substitute photo or signature. Advanced solutions may solve some but not all of these security weaknesses. Biometrics can be used to try to ensure a match between a document holder and their document ('authentication'), but even if the biometric authentication system is itself foolproof, for 'identification' purposes the system is of use only if the document is linked to a database of known identities, and even then is reliable only insofar as the registration process of establishing the document applicant's identity was itself reliable. A body may not lie, but its owner may previously have done so (cf. Franko Aas 2006).

Beyond the individual: other roles of security checkpoints

While some security checkpoints aim to establish a person's identity, or require them to authenticate, not all do. Three further common roles of security checkpoints can be identified, all of which are forms of 'authorisation'. First, and most simply, a checkpoint may be designed to check that a person holds a valid ticket for entry or travel. At one level, the 'security' thus afforded is weak to non-existent, since anyone with sufficient money can gain admittance. However, the 'security' (in its broadest sense) implications of ticketing can also easily be overlooked. Tickets typically authorise the holder to gain entry only on specific days or times, or are otherwise time-limited.

Moreover, ticket price levels can effectively limit ticket affordability. Where affordability maps to social class, ticket prices can be used to restrict entry to members of wealthier classes, whether to entertainment venues, transportation or particular residences.

Second, security checkpoints may be used to restrict admittance to certain 'types' of person, whether or not an additional ticketing system is in place. For example, admittance may be restricted to those of a certain age, degree of sobriety, gender, race or appearance (including attire). Such restriction may or may not be discriminatory, and may or may not be lawful, but does sometimes happen. Lastly, security checkpoints may be concerned primarily with an individual's possessions, whether in their baggage, within their clothing or otherwise about their person, or even concealed within their body itself. Prohibited items may include weapons, explosives, flammable materials, illegal drugs, recording devices, or bottles of water.

Each of these three roles is a form of authorisation, because only those possessing a valid ticket, or of a certain 'type', or (not) carrying certain possessions, are allowed to pass through the checkpoint. Given all the above roles that security checkpoints may play, and the kinds of selections and assessments they make, we can now consider the different kinds of ways in which they are accomplished (or their accomplishment is attempted, to varying degrees of success) in practice. Clearly, there are many ways in which one could go about trying to answer these questions, but in the following section I will introduce and discuss a fourfold typology as a way of developing the 'regulatory models of governance' approach suggested by Zedner (2006) among others.

The regulatory dimensions of security checkpoints

As I will try to show, gateways and security checkpoints can usefully be considered along a number of dimensions, sometimes harmonious, and sometimes conflicting. As I have argued elsewhere (Jones 2006; 2007), one starting point for modelling regulation in general has been suggested by Lessig (1999; 2006) in the context of his discussions of internet regulation. His model is intentionally general, though, and I have sought to show how it can be used within criminology in understanding various aspects of crime control and criminal justice. Lessig argues that there are four main ways in which regulation can be attempted: law, the market, social norms and 'architecture'. A figure from his book *Code* illustrates these four regulatory modes, with the dot in the middle representing whatever it is that the regulator seeks to regulate.

Offering the example of the dot representing a smoker, Lessig explains that one could seek to regulate their smoking by the law (for example, prohibiting smoking in enclosed public spaces), by the market (increasing the price of cigarettes), by social norms (such as by a public campaign encouraging

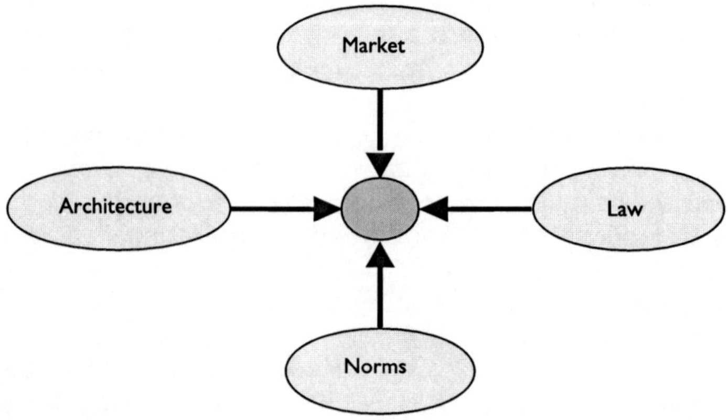

(Source: Lessig 1999: 88)

people to regard it as socially unacceptable to smoke in front of children) or by architecture (changing the design of a cigarette filter, for example, to alter the amount of nicotine inhaled) (1999: 87). Architecture can refer to the design not only of things but also of places, built architectures, environments, as well as the virtual environments of computers, their operating systems, applications and networks. Architectures establish environments in which certain activities are possible, while others are not. The four regulatory modes identified could be used separately, in concert, or indirectly on one another. Lessig emphasises that whatever the regulatory aim – and whatever the regulatory mode(s) employed – certain political values are always embedded therein. Indeed, one of the aims of Lessig's book is to show how different political values are reflected in architectural aspects of different computer networks, for example in the degree of anonymity afforded to users: a computer system can be designed to allow anonymous usage (and thus potentially enhance freedom of speech) but at the potential expense of security, or vice versa. However, whereas Lessig develops the model specifically in relation to regulation (principally by the government or regulatory authorities), I propose here to use the model to refer to 'social forces' more generally, by which I mean the range of imperatives impacting on or harnessed by people not only at a national level but also in relation to institutions, the private sector and individual interests, including where these various levels and interests are competing against one another. Clearly, Lessig's model is a highly simplified schematic, and I will use it simply as a useful heuristic to begin to identify some of the competing social forces at play in relation to security checkpoints.

The 'architecture' of security checkpoints

Lessig's model can usefully be applied to security checkpoints. At their simplest, gateways in general can be thought of as consisting of a barrier between two spatial regions (see also Franko Aas 2005), incorporating a managed gate of some sort, such as to allow some people to move from one side of the barrier to the other, while preventing others from doing so. At the heart of security checkpoints is the requirement that the barriers that stretch away on either side have physical integrity – otherwise people have no reason to submit to the checkpoint. Whereas a basic gateway can be opened or closed, letting all or none through, checkpoints typically require persons and their belongings to pass through a narrow channel for closer scrutiny. In addition, however, architectural and physical-compliance systems typically are used *within* checkpoint security regions so as to act upon their central targets of bodies and possessions, for example in channelling individuals (typically into single file) ahead of the checkpoint proper, or within a scanning or questioning process itself.

The significance of the use of barriers to manage crowds of people and prepare them for entry to transportation facilities and entertainment 'rides' was recognised by Shearing and Stenning (1985) in their study of Disney World. They argued that the use of barriers was an important emerging element in social control in public or quasi-public spaces. Their focus, however, was on social control and order maintenance, the barriers silently managing the crowds:

> Opportunities for disorder are minimized by constant instruction, by physical barriers which severely limit the choice of action available and by the surveillance of omnipresent employees who detect and rectify the slightest deviation.
>
> The vehicles that carry people between locations are an important component of the system of physical barriers. Throughout Disney World vehicles are used as barriers.
>
> (Shearing and Stenning 1985: 344)

To this we can add that these barriers can be used to move people along predefined transit routes, and to ensure they line up in an orderly fashion at ticketing booths, entrances to transport and entrances to rides. A feature of a relatively narrow corridor established by barriers of some kind (which might be established by means of waist-high metal bars, velvet ropes or tensioned belts) is that it forces crowds into a line of smaller family/friend groups, or even to separate into individuals. As these persons progress to the front of the line, they can then be processed one by one.

In the case of an amusement park such as Disney World, the security processes involved are likely to be ones of authorisation issuance and checking, rather than of identity or even authentication checking. Similarly, a leisure facility such as a theatre may use a gateway of some sort; patrons may be required to purchase a ticket for presentation to an usher who then allows them entry to the theatre proper. A more automated form of ticketing gateway can be found at many rail and underground train stations, with barriers through which one can pass if one has a valid ticket or electronic pass. Such ticketing gateways are essentially part of an authorisation system, the ticket itself being a kind of authorisation token.

Scanners: from airport security to airport-style security

'Airport security' can involve a range of security matters, each requiring a specific security approach, including airplane security, health, customs, and the national and regional security of border controls. Architecture can be employed in airport security checkpoints in several ways. One of the purposes of airport check-in is to establish that the traveller has a valid *ticket* for travel, but other practices carried out there include checking that the passenger is *authorised* to travel (carries a valid passport and, if necessary, visa) and to *authenticate* (typically by comparing the face of the traveller against the photo in their passport or ID card) that they are who they claim to be. Additionally, however, a process of *identification* may be conducted as part of the checkpoint security process, formerly against a printed sheet of 'wanted' persons, but today typically against a remote database. As Lyon (2006) notes, at Canadian airports schemes such as the 'Advanced Passenger Information (API) and the Passenger Name Record (PNR)' schemes have been introduced as a 'means of tracking travellers', and specifically to establish the identities of passengers before, during and after their flights.

Some airports have recently experimented with the introduction of biometric authentication systems. Biometric measurement variables can include 'Fingerprint, facial recognitions, voice recognition, iris scan, retina scan, hand geometry, [and] signature scan' (Wells and Rodrigues 2004: 324), but as Prabhakar et al (2003) and Dror (2006) note, these new systems are not infallible. In particular, given that such systems rely on sampling techniques rather than definitive judgement, they necessarily make errors, either of 'false positives' (here, accepting someone who should not be accepted) or 'false negatives' (rejecting someone who is in fact legitimate). Typically, a balance must be struck between the two kinds of error; where it is struck is likely to depend on social determinations of potential consequences of different kinds of mistake. Biometric systems could also, or alternatively, be used in identification systems, but this requires the establishing of very large (national or international) databases, matching people's records with their biometric measures.

Airports are designed to flow people through from check-in to security and to boarding, and from disembarkation to immigration, customs and the freedom beyond. (Some airports achieve this more efficiently than others.) Passengers and their baggage are processed by means of queues and corrals, and barriers and gateways are used to prohibit unauthorised movements between different zones of the airport. (The use of physical barriers and secured entry may continue within the aircraft itself in the form of cockpit door reinforcement, designed to 'protect cockpits from intrusion and small-arms fire' (Wells and Rodrigues 2004: 327).) By use of tickets, travel documents, boarding passes, baggage tags and baggage seals, people and their baggage are given authorisation to move from one stage to another. At security checkpoints, further physical properties of people and things are exploited, with the aim of policing compliance with specific airline or airport security requirements. These may include using scanners to detect compliance with restrictions on the amount of liquid that can be carried in hand luggage, or with prohibitions against carrying items such as explosives, toxic substances, radioactive materials, gas cylinders or infectious substances. In the context of a discussion of aviation and airport security in relation to terrorism and safety concerns, Sweet (2004: Ch. 7) characterises security screening as 'the last line of defense' (subsequent to intelligence and policing strategies, for example). However, in relation to policing compliance with restricted item requirements, in practice such screening may amount to being the first and indeed only line of defence.

Scanning and detection systems

One of the ways in which people and their belongings can be checked is by being physically searched by security guards. This is relatively slow, potentially physically intrusive, and potentially dangerous for those doing the searching. Security scanners, such as the metal detecting 'security arches' found at airports, offer a swifter means of scanning people for metal objects, and more sophisticated x-ray machines can be used by trained security staff relatively quickly to examine the contents of bags without needing to open them.

There are various kinds of scanning and detection technologies. The first kind is based on 'imaging technologies', which as the term suggests generate a special image of the person or object being scanned, typically enabling the scanner to see something which could not be seen using the naked eye alone. Imaging technologies:

> work either by sensing the natural radiation emitted by the human body (passive imaging) or by exposing subjects to a specific type of radiation and then measuring the radiation reflected by the body (active imaging). These systems can detect metallic weapons or plastic explosives by sensing the differences in reflected radiation between the human body and the

> weapons or explosives. The screening systems then generate televisionlike (sic) digital images.
>
> (Wells and Rodrigues 2004: 316)

Imaging technologies include x-ray machines, which today are available with computer-assisted displays, featuring colour coding of different types of material or even of automatic object outline recognition, and active millimetre-wave imaging (also known as 'backscatter x-ray imaging'), which uses 'low intensity reflected x-rays to scan an object' or person, effectively allowing the operator to see beneath a person's clothes (Wells and Rodrigues 2004: 317).

A second genre of scanning and detection system employs 'trace detection technologies', which 'are based on the direct chemical identification of either particles of explosive material or vapor-containing explosive material' (Wells and Rodrigues 2004: 318). These 'sniff' the air near to an object, or sniff for particles that have lodged on material. A third type of detection system are 'bulk explosives detection systems' (EDSs), which attempt to scan baggage or containers, often using x-rays, in order to determine the density of the contents, looking for substances that match the properties of known explosives, while a fourth type of system are metal detectors, which can take a number of forms. One of the most commonly used today is the 'portal-type metal detector', but handheld metal detectors are also used to pinpoint the exact location of items that have registered on a portal detector. Among the various other types of metal detector available is the chair-like 'body orifice security scanner' (BOSS), 'designed to detect metal objects hidden in body cavities' (Wells and Rodrigues 2004: 323).

Some of the security practices discussed above have now spread beyond the confines of the airport and can be found employed in various other locations (including courts, prisons and schools), where they are often colloquially referred to as 'airport-style security', suggesting that these practices may now have taken on a certain cultural life of their own. While Simon (2007: 208) cites survey research suggesting that only 1 per cent of US schools 'routinely screened students with a metal detector', the survey in question dates from 1996–7, the proportion still equates to a significant number of schools, and this number may well be higher today. As Simon also notes, US surveys have suggested that 'fortress tactics' in general 'including mandatory drug testing, metal detectors, and searches are hardly confined to a handful of the most crime ridden schools in America'. It is possible there is a cultural readiness on the part of some schools to implement checkpoint security were it to become economically and practically feasible. It seems plans are currently being considered for the introduction of airport-style metal detectors in some schools in England. Indeed, the installation of metal detector arches is just one way in which gateways can be converted into security checkpoints; in a discussion about letter bombs and security, Neyland (2008) reproduces guidance from the UK Security Service which essentially amounts to guidance on how to convert a mail room into a security checkpoint, through the introduction of scanners, practices and plans.

Traditionally, security checkpoints have been manned. However, an interesting aspect of increased reliance on architecture within security checkpoint systems in the real world is that this can be harnessed to remove attendant human staffing, for example by using electronic access controls to police entry (see for example Jones 2000; Lianos and Douglas 2000). Indeed, we can regard electronically controlled access systems as a special case of security checkpoint. Depending on the technologies employed, these electronic checkpoints may be able to authenticate (for example, biometrically), identify (by querying such data against a remote database), as well as check authorisation. One of the implications of such a development is the shifting of discretionary judgement from the checkpoint staff (cf. Skolnick 1975) to those in control of a remote computer system (cf. Lyon 2003b).

One of the intriguing aspects of 'architectural' measures is the apparent totalising quality of the regime they bring into being: there appears to be no arguing with walls, barriers or other 'designed in' systems. However, in fact, no security system is perfect, and indeed any architectural system can be subverted, more or less easily. Conversely, its 'success' may derive less from physical imperatives and more from surrounding social factors. Having examined some aspects of the 'architecture' of checkpoint security, I will now turn to a consideration of each of Lessig's other three dynamic modes in turn, suggesting in particular how these may either underpin or potentially confound the security the architectural techniques might be supposed to achieve. In this way, I hope to present a more 'realistic' model as to the strengths but also limits of checkpoint security systems.

Law

Restrictions on people and things

Over the past decade, European countries have witnessed a reconfiguration of border checkpoint security aims and purposes in response to European law and policy, with customs controls becoming less prominent, passport controls bifurcating between European and non-European passport holders, and tighter security of Europe's external border. Law can also be used to prohibit the moving of certain items through checkpoints. In relation to airplane security, for example, Neyland (2008) notes how legal rules about carrying sharp objects and liquids 're-oriented' security practices at airports.

Legal powers of checkpoint staff

Where security checkpoints are manned, their personnel (ranging from police officers, prison officers, airport security personnel, immigration officers, store detectives, to bouncers) possess varying legal powers. As Button notes, private security officers in England and Wales generally have no special powers over and above those of any citizen. 'They have no grounds to force a search', for example, and while they may ask a suspected person if they can look in the person's bag, the person need not agree (2007: 31–2). Exceptions include where searches are expressly made conditions of employment, as for example for some Sainsbury's supermarket staff (Button 2007: 39–40). Security officers working at ports and airports, however, are among those who have 'special legal tools under specific legislation in England and Wales'. These search powers could presumably be exercised by the officers in various parts of the institution or place in which they work; however, in practice these powers are likely to be exercised at a security checkpoint, through which all visitors and staff are expected to pass.

While '[t]here is no legal right for a security officer to forcibly undertake a search of the person or belongings of an individual seeking entrance to private property, . . . [o]wners of private property, however, can insist upon a search as a condition of entrance' (Button 2007: 38). 'The security officer must gain the consent of the individual to search their person and/or belongings', and if the person refuses 'permission they cannot be forced to undergo a search. Nevertheless if permission to search is refused the security officer can deny them entrance'. As Button notes, 'Clearly this is a very strong incentive for an individual to submit to a search'. Moreover, while a search can be requested once a person has entered a property, and 'they can be asked to leave' if they decline, such 'a sanction may not be as strong as on entrance', since 'the individual might have enjoyed their time there already, refuse, and therefore accept removal' (2007: 38–9).

Social norms

Social norms on the part of the security staff involved as well as of those passing through are likely to inform gate practices, the degree of security achieved and how this is to be understood. Practices may be conducted in the name of 'security', but actually be because of a desire to socially sort, segregate, achieve status distinctions or appeal to a certain clientele. Nightclub or restaurant doorstaff may do this too, or they may follow certain 'rules of thumb', such as was the case with some of the doorstaff interviewed by Hobbs et al (2003): 'If they're aggressive with me at the door, then they'll be aggressive inside . . . So don't let them in' (2003: 120) or people wearing certain clothing or footwear may be barred. This could be happening because to the doorstaff such appearance or behaviour symbolises potential trouble or because of house 'rules' in search of conferring social distinction on the establishment and its patrons, or it could simply be discriminatory or even entirely arbitrary (see Hobbs et al 2003).

In their study of gated communities in the United States, Blakely and Snyder distinguish between three types of communities, namely 'lifestyle communities, prestige communities, and security zone communities', 'represent[ing] differing physical characteristics and differing motivations of their residents' (1997: 38–39). Only the latter of the three types, they suggest, was strongly motivated by its residents' fear of crime; lifestyle communities offer 'security and separation for the leisure activities and amenities offered within', while the gates of prestige communities 'symbolize distinction and prestige and create and protect a secure place on the social ladder' (1997: 39–41). In one community they visited, they report that the residents played down the security benefits of the gates; their gate guards are 'relaxed, really only monitoring and slowing traffic, and no one is willing to pay the cost for improved levels of security. Like nongated communities, the neighbourhood has burglaries and other mild vandalism. "We live in a traffic-controlled community, not a secure community," says Jim' (1997: 88–89).

As Bourdieu (1994) explains, people's lifestyles, including their tastes in living places, food, clothing, and eating places, are among the ways through which social tastes and hence social distinctions are expressed and reiterated. 'Security' may be part of this process both in achieving a form of social sorting and in the symbolic role of the checkpoint itself. *How* the security is conducted is important in this respect too; Hobbs et al recount how 'the owner of an independent "upmarket" bar and restaurant' was careful to hire doorstaff who would project a certain image and be able to interact with patrons in a certain way (2003: 136–7).

Norms as to what a security checkpoint stands for, how seriously it is to be taken and how those passing through can expect to be treated can be expressed through the symbolism of the checkpoint's architecture, design and décor. In relation to prison architecture, Garland draws attention to historical research on the symbolism of 'prison façades, portals, and entrance lodges', and argues that whereas the exteriors of prisons today:

> are generally designed to serve the ends of security, containment, and anonymity, rather than deliberate or carefully construed representation . . . these muted, functional buildings nevertheless project an eloquent and well-understood symbolism, which speaks of unshakeable authority, of stored-up power, and of a silent, brooding capacity to control intransigence.
>
> (1990: 258–260)

The symbolism of security checkpoints seems to range across a fairly wide spectrum, from the quiet, bureaucratic intimidation of passport control checkpoints to the techno-functionalism of security tag scanners surrounding shop doors, but each connotes a certain 'imagery' of control to their audiences (see also Garland 1990: Ch. 11), informs social understandings of security checkpoints, inflects the social norms surrounding their use and brings particular 'cultural effects' (see Garland 2001: 163, *passim*).

Markets

Economics, and in particular financial self-interest, can act as a powerful inducement to compliance with checkpoint security practices, but conversely, the lure of extraordinary profits to be made from smuggling is a potential major source of checkpoint corruption. Indeed, there are various ways in which markets and economic considerations appear as major factors influencing the operation of checkpoints.

Ticketing security: ticket holders only

Checkpoint security may directly involve a ticketing system, such as when one pays admission at a turnstile and is immediately granted access, or may run in parallel to a ticketing system. An example of the latter can be found at airports, where passengers encounter multiple checkpoints or checkpoint processes, some of which are designed to authenticate a valid ticket for travel, while others are designed to verify the traveller's identification, and yet others are designed to scan the passenger and their belongings for dangerous items. However paid ticketing is implemented, it potentially serves to underpin compliance with the architectural aspects of checkpoint security, since the person seeking admission has already made a financial investment in the form of their ticket. This may be one of the reasons travellers are relatively willing to put up with security checkpoint delays at any given moment: they stand to lose the value of their ticket if they decide not to submit to the checkpoint process. However, for the same reason, lengthy delays or unwarranted invasions of privacy (from the traveller's perspective) may dissuade them from travelling via that route in the future (Ito and Lee 2005). Aware of this problem, both the air travel and airport retail industries have a significant incentive to try to streamline the airport checkpoint security process, or even to press for lower levels of security than are preferred by the government of the day (see also Neyland 2008).

Privatisation: weakened involvement

A second way in which economics could theoretically factor in checkpoint security effectiveness is in relation to staff employment practices. Zedner (2006), for example, draws attention to the high staff turnover in the private security industry in general, indicative of poor wages or working conditions. Low wages and high staff turnover are likely in turn to result in lower levels of training, expertise and commitment than would otherwise be the case. Lippert and O'Connor argue that 'physical-security provision is a form of work carried out by a workforce. In this regard, security entails "engaging" a workforce to act on the conduct of others. The nature of this engagement is a significant element' in the 'security assemblage' to have emerged in response to global insecurities (2003: 334; 350). They suggest that '[t]he insecurity of global capitalism has been for the most part shifted onto the lower echelons of the non-standard working class' and that the introduction of flexible (but also low-paid) working practices in response to the dynamics of global insecurities has contributed to 'a qualitative decline in the guardianship of those deployed within the airport security assemblage' (2003: 350–51).

Corruption

A separate but potentially related problem is that of security checkpoint staff corruption. The low wages offered by private security firms whose staff are manning checkpoints may make it easier to bribe the staff, but this problem would seem not to be limited to the private sector. Low pay is also to be found in many lower-grade civil service positions around the world. Indeed, the relationship between corruption and staff role may go further. As Zedner notes, one of the attractions of otherwise low-paid positions as nightclub doorstaff is the access to illegal markets the job enables (2006: 272). It may even be in some places that the income available from bribes or illegal trades becomes normalised and widely accepted.

In an ethnographic study of illegal trade flowing across several continents, Nordstrom identified corrupt customs officials as just one group colluding in an apparently normalised set of illegal trading practices: 'All these people know that economics is a dance of the il/legal: a pas de deux' (2007: 206). Similarly, van Schendel (2005) discusses how illegal flows can become 'domesticated' by people living near or working within borderland regions, since:

> a heavily guarded segment of the border can easily be a segment where border guards are heavily involved in private gain from cross-border trade. . . . If the evidence is to be believed, the very sentinels of the state are often highly susceptible to the lure of the borderland and become active agents in forms of scalar structuration that weaken state territoriality and strengthen illegal flows.
>
> (2005: 58)

Corruption may be present wherever money is to be made from moving valuable people or things across borders. The availability of drugs and other contraband inside prisons probably also sometimes involves prison staff (though not necessarily prison guards) to some extent (see, for example, Crewe 2005: 464–5). Whatever the security checkpoint concerned, whether at a local or international level, the point here is that the huge profits that can be made from smuggling need to be recognised as an important factor to be taken into account when examining the dynamics of security checkpoints.

CONCLUSION

Whereas Surveillance Studies has correctly and usefully pointed to how globalised surveillance has led to a certain 'delocalisation of the border' (Lyon 2003; cited in Franko Aas 2005: 208; see also Lyon 2005), in this chapter I have sought to show how checkpoint security at physical borders of various kinds is also important. Focusing specifically on the category of security checkpoints, I suggested that Lessig's regulatory schema could be useful as a heuristic for disaggregating some of the wider dynamics at play.

No security system can ever be completely secure, but architecture is likely to be favoured in the ongoing social project of 'designing out' security flaws. As situational crime prevention maintains, instead of counter-factually invalidating its approach, security breaches can be studied and learned from, and security measures revised accordingly. However, architecture alone can deliver only so much security. An assumption that all participants necessarily want improved security may not be correct: government and management may favour it, but staff and customers may not. Architectural security may be subverted for a variety of reasons, as was outlined above, and moreover even from within. One solution may be to try to securitise security measures and staff themselves (second-order security). But whilst this might itself be achieved architecturally, understanding why and how to implement such security checks on security checkpoints would seem to require understanding of the various legal, normative and economic forces at play.

I have emphasised how security checkpoints can effectively operate on a 'standalone' basis, subject to specific internal mechanisms and dynamics. However, it is also apparent that these otherwise isolated flow-control points can be linked together, or to remote data archives, using technologies old or new, and hence to be co-opted as part of wider surveillance practices. Lyon (2003b) has suggested an interesting way of thinking about centralised processing of electronic data, namely as a 'social sorting'. Checkpoint security appears a 'real-world' equivalent of such practices, sorting and checking individuals, granting access (and hence social benefit) to some but not to others, and policing how this access is permitted. There are, though, both potential further linkages but also differences between electronic social sorting and the sorting conducted by checkpoint security. One such difference is that whereas the challenge of electronic social sorting derives from the centralising, totalising quality of its rule application, a traditional problem with security checkpoints is that their isolation and autonomy potentially allow them to undertake arbitrary and unaccountable sorting. Such isolation seems likely to diminish as processes of globalisation and interconnectedness mean that forces of architecture, law, social norms and the market link previously unconnected social spaces. As a result, we may see ever-closer ties forming between physical security checkpoints and remote data systems. One use of such systems could be to gather surveillance on persons passing through checkpoints, but another is that data is increasingly likely to be used in deciding whether or not to permit someone through – a very immediate and real form of social sorting.

Security checkpoints riddle societies. From libraries to prisons, from law courts to border posts, security checkpoints filter the flows of people passing through. It is up to us to decide how far we want to rely on the technologies to assuage our insecurities.

References

Adey, P. (2006) 'Divided we move: The dromologies of airport security and surveillance', in T. Monahan (ed.) *Surveillance and Society: Technological politics and power in everyday life*, London: Routledge.

Adey, P. (2004a) 'Secured and sorted mobilities: Examples from the airport', *Surveillance and Society*, 1(4): 500–19.

Adey, P. (2004b) 'Surveillance at the airport: Surveilling mobility/mobilising surveillance', *Environment and Planning A*, 36(8): 1365–1380.

Auge, M. (1995) *Non-Places*, London: Verso.

Blakely, E. and Snyder, M. (1997) *Fortress America: Gated communities in the United States*, Washington, DC: The Brookings Institution.

Bottoms, A.E. (2001) 'Compliance and community penalties', in A.E. Bottoms, L. Gelsthorpe and S. Rex (eds) *Community Penalties: Change and challenges*, Cullompton: Willan, 87–116.

Bourdieu, P. (1994) *Distinction: A social critique of the judgement of taste*, London: Routledge.

Button, M. (2007) *Security Officers and Policing: Powers, culture and control in the governance of private space*, Aldershot: Ashgate.

Castells, M. (2000) *The Rise of the Network Society*, Oxford: Blackwell, 2nd edition.

Castells, M. (1989) *The Informational City*, Oxford: Basil Blackwell.

Cayhan, A. (2005) 'Policing by dossier: Identification and surveillance an era of uncertainty and fear.' In Dider Bigo and Elspeth Guilds (eds), Controlling frontiers: Free movement into and within Europe. Aldershot, U.K.: Ashgate.

Cornish, D. and Clarke, R.V. (2003) 'A reply to Wortley's critique of situational crime prevention', in M. Smith and D. Cornish (eds) *Theory for Practice in Situational Crime Prevention*, Cullompton: Wiley.

Crewe, B. (2005) 'Prisoner society in the era of hard drugs', *Punishment & Society*, 7(4): 457–481.

Dror, E. (2006) 'Cognitive science serving security: Assuring useable and efficient biometric and technological solutions', *Aviation Security International*, 12(3): 21–28.

Ekblom, P. (1999) 'Can we make crime prevention adaptive by learning from other evolutionary struggles?', *Studies on Crime and Crime Prevention*, 8(1): 27–51.

Florida, R. (2005) *The Flight of the Creative Class*, New York: HarperCollins.

Florida, R. (2004) *Cities and the Creative Class*, London: Routledge.

Foucault, M. (2007) *Security, Territory, Population: Lectures at the Collège de France 1977–1978*, Basingstoke: Palgrave Macmillan.

Foucault, M. (1979) *Discipline and Punish*, Harmondsworth: Penguin.

Franko Aas, K. (2007a) 'Analysing a world in motion: Global flows meet "criminology of the other" ', *Theoretical Criminology*, 11(2): 283–303.

Franko Aas, K. (2007b) *Globalisation and Crime*, London: SAGE.

Franko Aas, K. (2006) ' "The body does not lie": Identity, risk and trust in technoculture', *Crime, Media, Culture*, 2(2): 143–158.

Franko Aas, K. (2005) ' "Getting ahead of the game": Border technologies and the changing space of governance', in E. Zureik and M. Salter (eds) *Global Surveillance and Policing: Borders, security, identity*, Cullompton: Willan.

Garland, D. (2001) *The Culture of Control: Crime and social order in contemporary society*, Oxford: Oxford University Press.

Garland, D. (1990) *Punishment and Modern Society*, Oxford: Clarendon Press.

Hayward, K. (2004) *City Limits*, London: Cavendish Glasshouse.

Hobbs, D. et al (2003) *Bouncers: Violence and governance in the night-time economy*, Oxford: Clarendon.

Ito, H. and Lee, D. (2005) 'Assessing the impact of the September 11 terrorist attacks on U.S. airline demand', *Journal of Economics and Business*, 57(1): 75–95.

Jones, R. (2007) 'The architecture of policing', in A. Henry and D.J. Smith (eds) *Transformations of Policing*, Aldershot: Ashgate, 169–190.

Jones, R. (2006) ' "Architecture", social regulation, and situational punishment', in S. Armstrong and L. McAra (eds) *Perspectives on Punishment: The contours of control*, Oxford: Oxford University Press.

Jones, R. (2000) 'Digital rule', *Punishment and Society*, 2(1): 5–22.

Lessig, L. (2006) *Code: Version 2.0*, New York: Basic Books.

Lessig, L. (1999) *Code: And other laws in cyberspace*, New York: Basic Books.

Lippert, R. and O'Connor, D. (2003) 'Security assemblages: airport security, flexible work, and liberal governance', *Alternatives*, 28(3): 331–358.

Lianos, M. and Douglas, M. (2000) 'Dangerization and the end of deviance', *British Journal of Criminology*, 40(2): 261–278.

Lyon, D. (2007) *Surveillance Studies*, Cambridge: Polity.

Lyon, D. (2006) 'Airport screening, surveillance and social sorting: Canadian responses to 9/11 in context', *Canadian Journal of Criminology and Criminal Justice*, 48(3): 397–411.

Lyon, D. (2005) 'The border is everywhere: ID cards, surveillance and the other', in E. Zureik and M. Salter (eds) *Global Surveillance and Policing: Borders, security, identity*, Cullompton: Willan.

Lyon, D. (2003a) *Surveillance after September 11*, Cambridge: Polity.

Lyon, D. (ed) (2003b) *Surveillance as Social Sorting*, London: Routledge.

Neyland, D. (2008) 'Mundane terror and the threat of everyday objects', this volume.

Nordstrom, C. (2007) *Global Outlaws: Crime, money and power in the contemporary world*, London: University of California Press.
O'Malley, P. (2006) 'Risks ethics and airport security', *Canadian Journal of Criminology and Criminal Justice*, 48(3): 413–421.
Prabhakar, S. et al (2003) 'Biometric recognition: Security and privacy concerns', *IEEE Security & Privacy Magazine*, 1(2): 33–42.
Schneier, B. (2006) *Beyond Fear: Thinking sensibly about security in an uncertain world*, New York: Copernicus Books.
Shearing, C. and Stenning, P. (1985) 'From the panopticon to Disney World: The development of discipline', in A.N. Doob and E.L. Greenspan (eds) *Perspectives in Criminal Law*, Aurora: Canada Law Book Co., 335–349.
Simon, J. (2007) *Governing Through Crime*, New York: Oxford University Press.
Skolnick, J. (1975) *Justice Without Trial*, New York: Wiley.
Sweet, K. (2004) *Aviation and Airport Security*, Upper Saddle River, NJ: Pearson Prentice Hall.
Torpey, J. (2000) *The Invention of the Passport: Surveillance, citizenship and the state*, Cambridge: Cambridge University Press.
van Schendel, W. (2005) 'Spaces of engagement: How borderlands, illicit flows, and territorial states interlock', in van Schendel, W. and I. Abraham (eds) *Illicit Flows and Criminal Things: States, borders, and the other side of globalisation*, Bloomington, IL: Indiana University Press.
Wells, A. and Rodrigues, C. (2004) *Commercial Aviation Safety*, New York: McGraw-Hill, 4th edition.
Young, J. (2007) *The Vertigo of Late Modernity*, London: SAGE.
Young, J. (1999) *The Exclusive Society*, London: SAGE.
Zedner, L. (2006) 'Liquid security: Managing the market for crime control', *Criminology & Criminal Justice*, 6(3): 267–288.

Part III

(In)secure visibilities

Chapter 5

24/7/365

Mobility, locatability and the satellite tracking of offenders

Mike Nellis

INTRODUCTION

The satellite tracking of offenders' whereabouts on a daily basis, using the Global Positioning System (GPS) (latterly augmented by the Global System for Mobile communications (GSM)), has grown rapidly in the USA since 1997 and has since been piloted in a number of European jurisdictions (Bavaria, England, France and the Netherlands – see Elzinga and Nijboer 2006, Miedema and Post 2006) and also New Zealand. Continuing governmental interest in it beyond these jurisdictions suggests that geolocation technologies – developed originally to gain military advantage in the Cold War and to improve the safety and efficiency of global transportation systems – are now being perceived as potentially useful means of extending and modulating the intensity of state control over the everyday lives – the schedules, movements and locations – of at least some offenders. The commercial organisations which respond to (and stimulate) governmental concern with security are becoming increasingly adept at presenting satellite tracking as both superior to existing forms of electronic monitoring (EM) (which mostly monitor location in a single place, using radio-frequency technology, not satellites) and indispenable to the penal challenges being faced in the late modern western societies (Nellis 2008). Nonetheless, despite its expansion in the USA, it would be an extrapolation too far, at the present time, to claim that the future of satellite tracking is assured in Europe. Inferences might be drawn, however, from the emerging sociologies of surveillance and mobility that, in the medium term at least, it will expand further on the international scene.

Drawing selectively on a pilot scheme which ran in England and Wales between September 2004 and June 2006,[1] this chapter explores

1 I began work on the Home Office evaluation while working at the University of Birmingham, but left the project midway though the pilot in order to take up a post in Scotland. Several practitioners associated with the pilots kindly read earlier drafts of this article, and I would particularly like to thank the research associates on the project, Katherine Auty, Emily Evans and Katie Semro, for their detailed comments, as well as colleagues in the Ministry of Justice. The views expressed here are personal, while the official summary of the evaluation was written by Shute (2007).

the development of satellite tracking and appraises its significance as a new form of offender management, and its implications more generally for surveillance in contemporary societies. It is important to understand from the outset that 'satellite tracking', like EM in general, is not *merely* hardware but a socially embedded technology which is operationalised and experienced in a particular political, organisational and discursive context. It should not be assumed that insights garnered in one country will be exactly replicated in others, least of all until there is a common theoretical framework for understanding the distinctive forms of control that it entails. A preliminary sketch of such a theoretical framework follows below.

TOWARDS A THEORETICAL PERSPECTIVE ON SATELLITE TRACKING

The surveillance of mobility – the ubiquitous mobilities of people, artefacts and information – has become an established area of sociological enquiry, with a wide-ranging focus (Lyon 2002; Bennett and Regan 2004; Scheller 2004; Molz 2006). As yet, few theoretical insights from 'mobility studies' have percolated into criminology, possibly because there has been only limited criminological interest in offenders 'on the move' (although see Aas 2007). The satellite tracking of offenders (a quite literal instance of the surveillance of mobility) thus seems very novel to criminology, and even more so to many professionals involved in traditional forms of offender supervision. It is indeed intrinsically different from, say, probation or community service, insofar as it involves the automated monitoring (and sometimes restriction) of an offender's routine and everyday movements in limited spatial settings (quite often poor neighbourhoods) and the transfer of digitised data, including maps, between a range of criminal justice agencies. It *individuates* – in the sense of focusing on the movements of a single, embodied human entity – but it does not *individualise* – in the sense of seeking to know a person's inner mental life or to understand (with a view to changing) behaviour, as probation officers seek to do. This very distinctiveness rightly suggests that satellite tracking cannot be understood simply as an isolated and self-contained development in criminal justice, nor debated as such; rather, it signals the emergence of entirely new ways of ordering social life, which may have profound implications for humanistic approaches to crime control.

In general, it reflects the availability 'of new technologies [which] reconfigure the relation between space, speed, time and distance and the will of the bureaucracies of control to use them at their maximum' (Bigo 2006: 49). In particular, it bears out Bennett and Regan's (2004: 450) observation that 'the spaces in which the surveillance of mobilities regularly occurs [has] expand[ed] beyond those that are arguably hubs of mobility, such as airports, and now extend[s] *to any space in which people, objects or words move*' (emphasis added). It makes use, to a greater or lesser degree, of technical systems which have been created to undertake military intelligence, transport control (vehicle, boat and plane tracking, and the as yet rare road tolling systems) and to facilitate cellphone communication, all of which locate and track mobile individuals, as, in different ways, do CCTV systems and the audit trails left by digitised financial transactions:

> With the surveillance of mobilities there is potentially no 'hiding'. There is no room to walk anonymously down a street, drive through a neighbourhood, or talk on the phone. All these movements and flows are subject to scrutiny, captured, stored, manipulated and subsequently used for purportedly benevolent or underhandedly sinister purposes. The objects we use (cars, phones, computers, electricity) in turn become tools for surveillance. Movement is not a means of evading surveillance but has become the object of surveillance.
>
> (Bennett and Regan 2004: 453)

These, of course, are the routine, largely unreflective experiences of ordinary citizens, whose immersion in such systems demonstrates a mix of casual assent, begrudging acquiescence and active desire. Given the affordances of the technologies available, and the likely future direction of such technologies, it was arguably only a matter of time before very specific and sophisticated forms of mobility monitoring were applied to categories of people about whom there was probable cause to be suspicious, frightened or hostile. As Hannam, Sheller and Urry (2006: 1) put it: 'Fear of illicit mobilities and their attendant security risks increasingly determine the logics of governance and liability protection within both the public and private sectors.' Thus, while many citizens will choose *voluntary locatability* for their own convenience and security, some citizens will have *enforced locatability* imposed on them, using variants of the very same technologies. Although the concept of tracking offenders by satellite crystallised among technocratically inclined policymakers desperate to transcend, or at least augment, the old humanistic forms of offender supervision, it is against the backcloth of an existing, multiple-use technological infrastructure that its emergence must first be understood.

Mobile communication and geolocation technologies enable connectivity across space in ways that produce a sense of human proximity without the element of physical presence that would once have been required; they facilitate 'new ways of organising the spatial scale and temporal rhythms of interaction' (Scheller 2004: 42). Within criminal justice, the spectrum of electronic monitoring technologies – house arrest/curfew tagging, voice verification and now satellite tracking – is just such means of connectivity and is aptly thought of as 'automated *socio*-technical systems' (Lianos and Douglas 2000) precisely because, despite being defined by the *technological* nature, a human element remains (at least for now). Much has been written about the way in which 'virtual communication' can sustain a sense of relationship, solidarity and community among spatially dispersed networks of people, but that is not what is at issue here. EM merely facilitates data-gathering *about* someone rather than knowledge *of* someone, and it entails a dyadic link between a single (or split) authority (law enforcement agency/monitoring centre) and a subject, rather than multiple links within a network. One of the paradoxes of satellite tracking offenders – given the vast global reach of GPS – is that the degree of spatial separation between authority and subject is rarely great: it is relatively local, parochial behaviours which are being monitored and regulated. While the monitoring centre itself may be hundreds of miles away from the monitored subject, police and probation officers involved in the broader supervision programme are likely to be in the same neighbourhood.

Virtual communication technologies have created 'economies of presence' (Mitchell 1999) in which the balance of physical co-presence and remote contact necessary to the accomplishment of a particular social task can now be subject to routine cost-benefit analysis. The emergence of EM, which is often justified by its low cost relative to imprisonment, strikingly illustrates the way in which 'economies of presence' are migrating from the commercial field where they originated directly to the offender supervision field – and thereby transforming what is meant by 'supervision'. The periodic co-presence of supervisor and supervisee was once integral to the very meaning of supervision; it was via their structured personal encounters (and sometimes through the relationship which grew between them) that an impact on behaviour was effected. The application of remote monitoring technologies to offender supervision has enlarged the *spatial range* over which supervisory influence can be exerted – even house arrest/curfew tagging added a surveillant means of gaining compliance with a court order or release licence to the incentive-based, trust-based and threat-based means of gaining compliance which have traditionally comprised the social work/law enforcement repertoire. But, even more importantly, remote monitoring technologies have extended the *temporal range* of supervision within a given 24-hour period. In the past, the most intensive forms of personalised, humanistic supervision have rarely been more than intermittent, daytime encounters, while curfew tagging added in an element of control over night-time activities only. Both approaches leave offenders with significant periods of time when they are outwith the oversight of supervisors, when their whereabouts are *unknown or uncertain*. It is the temporality of satellite tracking that most distinguishes it from humanistic and relational forms of offender supervision, because it seemingly makes possible *incessant oversight* – round-the-clock knowledge of an offender's location, in real time or (more usually) some approximation to it – that no personal supervisor could manage and that no traditionally oriented social work or law enforcement agency could afford. This quality of incessance has become, quite literally, a major selling point of satellite tracking.

SATELLITE TRACKING IN THE USA

From the inception of house arrest with electronic monitoring in 1982, there was always a constituency among technology manufacturers and correctional agencies in the USA who wanted to track movement in space rather than *merely* monitor presence at a single location, not least because this was the kind of approach that the original Schwitzgebel experiments – from which the first patents on EM technology derived – had taken in the 1960s (Gable and Gable 2007). Satellite tracking schemes eventually began in the USA in 1997, in Florida, Michigan and Pennsylvania. Florida's community control programme led the field in this technology, readily using it with serious offenders, especially sex offenders. Active (real-time) tracking was used first, and although passive tracking was introduced in mid-2002, 'active' was still the preferred option, and the most common approach, among probation officers in 2004. It nonetheless heightened public expectations of safety and a correspondingly greater agency liability when something went wrong. The systems included the creation of inclusion and exclusion zones, 'two way communication with the victim or the offender, location mapping for archives retrieval, immediate tamper notification and remote laptop tracking with a wireless modem for constant communication with the monitoring centre'. Schools, day-care centres and parks could be 'hot zoned' to exclude paedophiles, bars and convenience stores to exclude drink-related offenders. From 2003 Florida used Crime Trax software to cross-reference offender locations with crime incidents – and claimed that the technology had solved six crimes in six months. (Securicor 2004). Thirty-two states had GPS schemes (not necessarily state-wide) by 2004.

Whilst decisions to introduce satellite tracking schemes are political (and economic, spurred by the need to reduce burgeoning prison costs by facilitating structured early-release programmes), the role of commercial organisations in promoting electronic monitoring in general and satellite tracking in particular cannot be discounted. A global 'commercial-corrections complex', rooted in but not exclusive to the USA, has emerged out of the old security industry to encompass the building and running of private prisons and, more recently, the telecommunication applications underpinning EM (Lilly and Knepper 1993; Christie 2000). Some of the organisations involved are indigenous US companies, others are global corporations. Some simply sell monitoring technologies to law enforcement and correctional agencies, others sell more sophisticated computerised case-management packages which include the technologies. All emphasise the cutting-edge nature of their contributions to crime control, marketing their product as a modernising technology offering more total oversight of offenders to correctional agencies than has hitherto been possible. 'Leap into the future', invites Shadowtrack, whose programme 'couples interactive voice response technology and voice biometrics authentication to keep track of offenders via the most modern and flexible telephone solutions available'.

Those companies marketing tracking technologies emphasise the powerful symbolic appeal of *incessance* – knowing where offenders are in real or near-real time, over sustained periods. iSECUREtrac, for example, plays directly on the temporal limitations of night-time curfew tagging: after headlining with 'Do you know your offenders are compliant when they're way from home? We check every 10 seconds!' they follow through with 'iSECUREtrac GPS systems offer you the truth. You can hold your offenders accountable to the places they've been and the times they've been there, 24/7/365, anywhere in the world. Additionally GPS tracking systems can greatly increase your level of offender supervision without adding to officer workload. iSECUREtrac alone can provide you with location and compliance verification every 10 seconds, fastest violation reporting on the market, user-friendly, yet powerful, web-based software; proven GPS policies and best practice for agencies'. Marketing a case-management package, Syscon seeks directly to dispel anxieties about offender's night-time activities with an advert depicting a sleeping man, explaining the contented look on his face as follows: 'This probation officer is using Syscon's automated systems to manage his low risk caseload with a range of kiosk, voice recognition and GPS technologies handling report-ins, the collection of fines, fees and restitution, and secure monitoring – all wrapped up in a fully integrated system. Only Syscon can offer you the full service package from end to end. It is no wonder he sleeps easy' (in *Journal of Offender Monitoring*, 19: 2).

THE SATELLITE TRACKING PILOTS IN ENGLAND AND WALES

At the turn of the millennium, England and Wales had the most extensive EM scheme in Europe and the New Labour government was already aware of the potential for using satellite tracking as a means of monitoring exclusion zones, although not yet convinced that fully reliable technology was available. It had legislated for such tracking in 2000 (in the aftermath of the murder of eight-year-old Sarah Payne by a known paedophile) in anticipation that technology would become available in the near future. It was personally championed by the then Home Secretary, David Blunkett, who dubbed it – hyperbolically, conjuring a sense of it as incapacitative – a 'prison without walls'. The immediate catalyst for the pilots' development was the Correctional Services Review (Carter 2004, Nellis 2005), which sought to 'modernise' the probation service, subsuming it within the new National Offender Management Service, and to promote the increased privatisation of correctional services. 'Modernisation' largely meant increased managerialisation and the repudiation of traditional humanistic probation practices; crucially it also meant expanding the use of information and communication technologies in all areas and at all levels of criminal justice (Home Office 2004a, 2004b). The increased use of EM – both curfew tagging and satellite monitoring – was considered integral to the modernisation of community supervision and the latter was infused into a range of existing organisational, legislative and policy concerns, including the intensive supervision of young offenders, the management of persistent and prolific offenders (the 10 per cent of offenders who commit 50 per cent of the most serious crime (Probation Circular 41/2004)) and of sex offenders, and the protection of crime victims. Limited evaluative data was available from America (see Nellis 2004) and tracking was deemed compliant – by dint of being 'necessary and proportionate' – with existing administrative law and the Human Rights Act 1998.

Although the Correctional Services Review presupposed that satellite tracking already had an assured future in Britain, the official aims of the pilots were modestly phrased: 'to gain practical experience of tracking technology' and 'to introduce a new sentence – a stand alone exclusion order/requirement' (NOMS 2002 para 1). The pilots were to be administered by the existing EM contractors, G4S, Premier and Reliance, although recontracting mid-way though the pilot period left only G4S and Premier (later renamed Serco) in the field. Serco used iSECUREtrac technology and later Benefon technology; Securicor used STaR (Satellite Tracking and Reporting) technology developed by the Israeli company ElmoTech. There were three initial pilot sites – selected police subdivisions of the Greater Manchester and West Midlands conurbations, and the whole of the county of Hampshire (including the Isle of Wight) in the south of England. The intention was to track both sentenced offenders and offenders released on licence; young adult persistent and prolific offenders in all three sites, sex offenders in Manchester alone and young offenders under 18 in Hampshire alone.

For offenders sentenced to a community penalty, the technology was to be used only to monitor compliance with exclusion-zone perimeters, and was expected to be used in the context of intensive supervision programmes. With offenders released on licence, however, it could be used as a stand-alone measure to monitor their whereabouts in general, *in addition to* any exclusion (or other) conditions included in their licence. With the former category there was at the time no legal power to track whereabouts in general, and while the monitoring companies did gather information on such offenders' general whereabouts, legally they could do no more than inform supervising agencies of perimeter violations. Protocols were governed by Statement of

Operational Requirements and Service Processes (NOMS 2005a, 2005b). Some months into the pilot, the standard American terminology of 'active' and 'passive' was changed to the following:

> *Continuous monitoring*. The location of an offender is reported in real-time to the monitoring company control centre. This requires a constant signal, and can therefore be very expensive, and, despite an intention to do so, was never used in the English pilot.
>
> *Retrospective monitoring*. The movements of an offender are uploaded once a day (or more, depending on the equipment being used) from the wearable tracking device, and reviewed later, usually the morning after the night of the upload. The interval between location fixes can be varied, but can be as short as every minute. This approach was widely used.
>
> *Exclusion zone (formerly 'hybrid') monitoring*. This combines retrospective tracking (and intermittent uploads) with real-time alerts of any violations of the exclusion zone, and then continuous tracking – location fixing every 30 seconds – *within* the exclusion zone [NOMS 2005a: para 6.31]. This was sometimes used, least frequently in Manchester.

The technology

All tracking technology used in the pilots employed two-piece units – a tag on the ankle and a tracking 'box' worn at waist/belt level or carried in a bag. Tag and box communicated using radio frequency, to within 5 metres of each other – if they became separated, i.e. if the offender abandoned the box, the box recorded this. A docking device was also installed in the offender's home, wired to the landline telephone system and, to recharge the battery (a daily requirement), the mains electricity supply. The tracking 'box' received signals from orbiting GPS satellites and when in communication with 'at least four satellites its position can be calculated in three dimensions, and can give an accurate location to within 10 metres' (G4S 2005). Locations were recorded every few minutes, stored in the box and uploaded to the monitoring centre via the docking device. As expected, there were extended periods of signal loss and many instances of 'drift' – inaccurate pinpointing on maps – which might be fully avoided only by upgrades to, or replacements for, the GPS system itself.

The more sophisticated tracking equipment also used GSM (cellphone) technology, to augment location indoors (where GPS is limited), to upload data to the monitoring centre, as well as to (in some instances) communicate by text or voice with the offender. GSM identifies which cell site an offender is in, to within 500 metres in urban areas and within several kilometres in rural ones, depending on the number of 'cell towers' (phone masts) available. Although these kinds of tracker still required battery recharging, docking with a landline was not required; the box automatically uploaded location information via the cellphone network. With retrospective tracking, the upload interval could be every four hours, but real-time tracking required constant uploading/signalling. Even when using the General Packet Radio Service (GPRS)[2] such monitoring is expensive and can generate paralysingly large amounts of information, which – except in relation to real-time alerts – it is difficult for agencies to analyse quickly .

Monitoring companies could retain location data for a limited period, although any relevant agency – police, probation, Youth Justice Board, legal representatives and medical authorities – could apply for information, each case being looked at on its merits (NOMS 2005a: para 9.1). No automated cross-referencing of satellite tracking data with crime-incident data occurred during the pilots, although in the course of the pilots police forces did occasionally request location data in the belief that it would either incriminate or exonerate an already suspected offender.

Exclusion zones

Within the pilots satellite tracking was intimately connected with excluding offenders from specific places, and notwithstanding the limitations of the 'panopticon' concept for understanding contemporary forms of surveillance, there was an element, at neighbourhood level, of 'banopticon' about the strategy (Haggerty 2006; Bigo 2006). Although not without precedent in bail, sentencing and release licences – offenders can already be banned from pubs, shopping malls, town centres, sports stadiums, public parks and victims' homes – spatial exclusion has hitherto played only a limited part in community supervision, mostly because it has been seen as a rather negative measure, at odds with professional imperatives to reintegrate offenders, but also because it can, without surveillance technology, be hard to enforce. However, recent crime-prevention legislation, with its flagship 'antisocial behaviour' and 'dispersal' orders, has strongly affirmed exclusionary principles, and there was a sense in which satellite tracking was being used to bolster its credibility.

The new exclusion zones, envisaged as alternatives to custody (part prevention, part punishment), prohibited offenders from entering a specified place or area for a specified period of not more than two years, or three months for those under 16. Monitoring companies required 72 hours' notice to assess the

2 GPRS is a connect-from-anywhere wireless data service available with most GSM networks. It was developed with a wide range of enterprise and consumer applications in mind, not least mobile internet services, eg colour internet browsing, e-mail on the move, video-streaming, multimedia messages and location-based services. With GSM, a fee is paid *each time* a position within the cellular network is requested, whereas the cost of transmission through GPRS is based on a fixed price for a fixed amount of data, whether that comes en bloc or at one-minute intervals throughout the day. This infrastructure was tapped by the monitoring companies as the cheapest means of real-time tracking.

technical viability of any proposed exclusion zone, to identify signal black-spots and program the computers. Although an offender's consent to exclusion-and-tracking was not formally required, it was required of the parents of under 18 year olds and any affected householders. Offenders were to be given written statements of the place and period of exclusion, and maps which clearly marked out boundaries and access routes. Stand-alone exclusion was always an option, but it was mostly to 'be an element of a programme of intervention' (para 6: 12). The legislation assumed that exclusion would always be ordered with EM, otherwise it would be unenforceable. Guidance from NOMS strongly recommended that exclusion zones be combined with curfews, *practically* to charge the battery overnight and *symbolically* to achieve round-the-clock oversight. (NOMS 2005a: 13). Exclusion was to be 'specific and proportionate', drawing on pre-existing guidance about exclusion zones in licence conditions (see Probation Circular 28.2003). In addition, when 'the purpose of the zone is to prevent the offender coming into contact with a victim, careful work will have to be done with the victim in order to manage expectations and explain the implications' (NOMS 2005a: 6.14).

Complex protocols were set out in respect of exclusion zone violations and breach procedures, allocating different responsibilities to police, probation, the monitoring companies and the Home Office, depending on circumstances. Where offenders were being continuously tracked, the police could be informed immediately of a perimeter violation, enabling – in principle – a rapid response to any victim, although adequate intervention time sometimes required disproportionately large zones. Temporal violations could be ascertained with some precision – for instance, a first-time perimeter infraction that '*lasted less than a minute* and presented no danger to the victims' warranted only a warning (NOMS 2005a: para 6.34, emphasis added). Actually judging whether a perimeter had been breached, however, sometimes required complex assessments – 'based on a calculation of plot accuracy derived from a number of plots, satellites available, signal strength and speed of subject' (idem) – which only technical experts from the monitoring company could explain to a court. A 'margin of error' was acknowledged, and one of the few tracked offenders to be convicted of breach on satellite tracking evidence alone, allegedly for travelling in a stolen car, successfully appealed on the grounds that the system had not, in fact, accurately pinpointed him.

TIME, TECHNOMANAGERIALISM AND COMMUNITY SUPERVISION

In general, satellite tracking has emerged as one of many affordances created by broader developments in information, communication and geolocation technology. The *precise forms* it takes in particular countries, however, are shaped by specific crime-control discourses and the wider political stuctures in which they are embedded. In the case of England and Wales, these were the expansion of managerialism, the intensification of punitiveness, and the officially constructed 'failure' of existing humanistic approaches to the community supervision of offenders, the former being the most important. Above and beyond the way in which managerialism redefines standards of economy, efficiency and effectiveness, it also seeks to impose regimes of meticulous regulation, to accelerate response times and to respond flexibly to changing circumstances. Applied to organisations involved in the community supervision of offenders, these imperatives alone created expectations which could not readily be met by traditional – slower, less precisely focused, more relational – humanistic interventions, which could then be quite easily branded as inadequate means of protecting the public.

So constitutive have computerised forms of communication been to contemporary managerialism (see Aas 2005) that it is hardly an exaggeration to call it *technomanagerialism*, and once that is done the 'elective affinity' between management and electronic monitoring is more easily seen (Mainprize 1996). In England and Wales, under the New Labour government technomanagerialism has been intricately embedded in discourses of *modernisation*, across the public sector, and strikingly so in criminal justice. This then has the effect of rendering traditional (non-technical) humanistic interventions anachronistic – if not quite obsolete then certainly in need of subordination to *more modern* ways of doing things. Managerialist practices – packaged as intrinsically modern – also help to resolve the political dilemma of making community supervision more punitive without – discursively – sacrificing progressive credentials and succumbing to atavism, because it is in the nature of such practices, applied in a probation context, to impose tighter regulation over the temporal and spatial movements of offenders (as well as staff). Whilst managerialist discourses lack the visceral energy that invariably informs demands for increased punitiveness, their practical application to offenders can have the effect of making the supervisory experience more onerous. Much of the pressure for more intensive forms of community supervision that has developed in England and Wales over the past 30 years has emanated from managerialist agendas and mentalities (sometimes manifest, sometimes latent), whilst being discursively entwined in official documents with retributive and incapacitative desires (Nellis 2002).

From a surveillance of mobilities perspective the community supervision of offenders can be regarded as a 'time-space practice' whose form is affected by technological development. Such supervision has traditionally relied on – indeed been defined, rationalised and legitimated by – the co-presence of offender and supervisor, in the offender's home or the supervisor's office, or some other agreed location. Although the synchronising of co-presence had much to do with the efficient use of a busy supervisor's time, punctuality and timekeeping on the part of the (mobile) offender has always been regarded as a putative index of compliance, and instances (or accumulations) of unpunctuality could be sanctioned. Gauging the amount of contact time between offender and supervisor necessary to achieve a desired result – rehabilitation and the reduction of offending – was the subject of both practical reflection and evaluative research from the 1960s onwards. Latterly, since the 1980s, a key concern among politicians and policymakers has been the amount of unsupervised free time that remains available to the offender in traditional forms of humanistic community supervision (Nellis 2002). Extending temporal control has thus been integral to the process of intensifying community supervision, and it is in that context that EM – first curfew tagging, then satellite tracking – has 'logically' emerged.

There remain some interesting anomalies. Partly because the Home Office initially used EM as a way of threatening the probation service to modernise its own practice, and partly because EM is delivered outside the probation service, by the private sector, there has been a sense in England that EM and 'effective probation practice' have developed on parallel tracks (Nellis 1991, 2003). Curfew tagging was envisaged as something that would, in the main, stand alone, rather than being integrated with other, potentially rehabilitative, measures. Satellite tracking was conceived differently from the outset, as an element in a rehabilitative-and-control package. Whether this reflects Home Office recognition that even sophisticated forms of EM are of limited value on their own is moot; it does reflect a more general aspect of virtual connectivity, namely that it does not so much dispense with the need for co-presence as create new configurations of proximity. As Jain (2006: 57) puts it: 'Copresence has not completely been replaced by ICTs, rather ICTs configure new layers of connectivity and present new opportunities for managing copresence, where the individual is situated within diverse spatial relations.' Thus, while there was a degree of impersonality in the intensive supervision programmes to which the *adult* offenders were subject, they were not known merely as 'data-viduals', they were individualised by the police and probation officers supervising them. Nonetheless, we should not be complacent. Virilio's (1995: 57) view that 'with real-time technologies, real presence bites the dust' may not (yet?) apply in this context, but both the logic of technomanagerialism – the 'economies of presence' that emerge in 'automated sociotechnical systems' – and a nascent penal willingness to devolve elements of 'interaction' with offenders to machines (perhaps as a way of signifying scorn) – suggest that if humanistic practices in community supervision are to thrive in the future, they warrant robust defence in the here and now.

EXPERIENCING MOBILE SURVEILLANCE

To understand the significance of satellite tracking from the offenders' perspective it helps to know that mobility – being out of the home, getting around, servicing friendship networks – is a very salient dimension of young British people's lives (Henderson et al (2007: 101). Disadvantaged/working-class youngsters often lead spatially circumscribed, highly localised lives, but all today's young people seemingly desire voluntary locatability; mobile phone ownership among them has introduced the possibility of greater improvisation and last-minute scheduling of encounters than was possible for earlier generations (Jain 2006). It also helps to be cognisant of the emerging literature on offenders' spatial behaviour and 'journey to crime', some of which faintly recognises the potential of EM in general for impeding and disrupting it (Rengert 2004), and extensive use by police of crime mapping with GIS technology (Weisburd 2004). This knowledge was not designed in to the pilots and had not yet filtered down to the practitioners who administered the intensive supervision programmes in which tracking was set, but as more is learned about 'planning criminals, wandering criminals and haphazard criminals' (Elfers 2004), doubtless it will be. In each pilot area sophisticated police computers were usually used to identify the locations where individuals committed most of their offences, but the rationale of particular exclusion zones was often a mix of punishment (keeping offenders from places that they liked to frequent) and prevention (disrupting offending patterns).

Overall, 517 offenders were tracked during the pilots (Birkett, Ballard and Smith 2007). Too few domestic violence perpetrators were subject to it (n = 15) for much significant comment to be made. (One victim at least was utterly unconvinced that the technology made her safer, and a tracked perpetrator simply used a proxy to harass his former partner rather than doing it himself.) Some sexual and violent offenders (n = 73) were tracked in the later stages of the pilots, and some imaginative approaches were developed, e.g. prohibiting one man from going to all a city's parks but, to save expense, creating electronic exclusion zones around only three, albeit without telling him which ones. The majority were prolific and persistent offenders (n = 329). In the West Midlands, these people bore out an established insight from 'environmental criminology', namely that 'offenders will more frequently commit offences in the neighbourhoods around their own domiciles, since these are the areas they know best' (Rhodes and Conly 1981: 169). They routinely burgled their neighbours in order to fund drug habits. Many of them lived with their parents, or in streets or on estates close by, and drew support from extended family and friendship networks. Creating exclusion zones thus posed a problem – how to enable them to live at home and yet keep them from streets with which they were familiar? The solution was to create exclusion zones (sometimes more than one) around or adjacent to their homes and permit only one access route. This did often mean that they were prohibited from visiting relatives and girlfriends (although there was nothing to stop relatives visiting them). By and large, they felt their exclusion keenly and were often powerfully motivated by the promise of gradual relaxations of it – afternoon or weekend access to particular places or people within it – to conform with the rest of their supervision regime – reporting to probation offices and police stations, random drug testing, random home visits, strict curfews, drug treatment and work seeking.

The exclusion zones created for young people in Hampshire (n = 100) tended to include more public spaces than were used in the West Midlands – football grounds, seafronts, shopping malls, where offences had occurred. In the early days, some youth court magistrates were reluctant to impose tracking on younger teenagers, despite social work recommendations to do so, seemingly because of the offender's small stature relative to the bulk of the box. Some young offenders subject to tracking were already subject to spatially restrictive Anti-Social Behaviour Orders (ASBOs), several of which were made coterminous with exclusion zones, creating, in effect, the first electronically monitored ASBOs. Some football banning orders, not originally designed to be combined with EM, were also monitored by satellite tracking. The social workers involved were caring and helpful towards the young people and used their discretion constructively when carrying the box would have otherwise interfered with a supervisee's capacity to participate in other activities (such as swimming).

Persistent and prolific offenders' opinions varied as to the onerousness of exclusion-and-tracking. There was often resentment that they were subject to something so demanding and intrusive *after* having left prison, many claiming that they had already 'done their time'. (This was as true of those whose licences required them to live in a hostel as to those undergoing satellite tracking.) Some of those who had never consented in the first place failed to keep appointments when the monitoring company called at their home, others simply smashed or discarded the box. Some likened the exclusion/tracking experience to prison itself because they felt so rule-bound and because permission had constantly to be asked of the authorities (to take a bus whose route happened to cross the exclusion zone, for example). Those in Manchester using ElmoTech's STaR equipment resented the frequent texted demands to find a signal when the box light went out – it required them to leave their house or break off mid-conversation, and it evoked a feeling of being monitored in real-time. This was perhaps the nearest offenders came to experiencing tracking as *incessant oversight*. Few, however, had a vocabulary for conceptualising it as surveillant or themselves as objects of surveillance, and in the West Midlands, where the equipment did not initially have this notification facility, offenders experienced exclusion from a place on the ground more vividly and concretely than they experienced being 'watched' by eyes in the sky. Some were simply indifferent to the abstract fact that the monitoring company knew where they had been, others felt their privacy had been infringed if the authorities knew they had visited a friend.

A handful actually likened tracking to freedom because, despite all the constraints, it *was not* prison, although one felt so restricted in where he could travel that he opted to stay indoors more than he otherwise would have done.

Some tested the boundaries of the exclusion zone, entering it to see what the consequences would be, and some lied about where they had been, until they were challenged by maps pinpointing their trails. Most made efforts to hide 'the box' from strangers and to avoid conversations about it; some claimed that it affected their employment opportunities and some were denied access to nightclubs when doormen enquired what it was. Some inevitably re-offended – one was caught stealing 'live' on CCTV – and were apprehended independently of being tracked, although maps could have corroborated their presence at a crime scene.

Whilst satellite tracking *is* a surveillant mode of gaining compliance, the subject is not intended to be – indeed *cannot* be – wholly passive. For tracking to work effectively the offender has to interact with the equipment and act responsibly towards it (more so than with ordinary tagging). The monitoring companies speak, quite aptly, of *enrolling* offenders on EM programmes, and the pilot protocols stated clearly that 'the ability of the offender to understand and comply with the demands of satellite tracking is a factor which needs to be considered before tracking is put forward as a stand alone condition or in support of other conditions' (NOMS 2005a: para 7.16). The 'Notice of an exclusion order with an electronic monitoring requirement' (idem: Appendix F) given to offenders lists 16 requirements which they must acknowledge and comply with. Among them are 'I agree to carry a tracking device whenever I leave and am away from my place of residence; I agree that I will not allow a distance of more than 5 metres to come between the tracking device and myself; I agree to charge and/or dock the tracking equipment while I am at my place of residence and when the device battery level is on one bar. I understand that I must take reasonable care of and not remove or damage the electronic monitoring equipment, or tamper with its ability to track my location'. Offenders were (formally, at least) allowed a fortnight's leeway 'to learn how to manage the tracking equipment' (Appendix B, NOMS 2005b) but in the event there were lapses, occasions when offenders forgot to carry the box, and their potential status as violations had to be negotiated with police and probation supervisors.

Within an undoubtedly coercive framework, there are nonetheless strong elements here of what Adam Crawford (2003) has called the 'contractual governance' of individuals, or 'regulated self-regulation'. Offenders cannot simply submit unreflectively to satellite tracking – they have to embrace the demands the technology makes of them (and sometimes respond to texted instructions) – and in that sense a degree of trust is still being shown towards them (which can be, and sometimes was, abused, not least in the smashing of boxes and the disregard of boundaries and appointments). EM vendors rather ruefully acknowledge that current satellite tracking technology is very 'participant-dependent' in this respect; ideally, it is implied, the offender should have much less leeway. It is worth noting that Satellite Tracking of People Ltd (STOP) (an American company) markets the Bluetag *one-piece* tracking unit under the slogan 'Track the *body* not the *box* and improve community safety . . .' and describes it as 'the least participant dependent device available today' (advertisement, *Journal of Offender Monitoring*, 19: 2). The most obvious element of participant dependence in EM generally is the removability of the tag, which can easily be cut off, and some British newspapers have derided this as a system-design flaw so great that it robs tagging (and by implication tracking) of credibility as a means of control. To the extent that there is a crime-control logic to reducing participant dependence, it seems plausible to suggest that sooner or later non-removable tags will come to be seen as vital to political and public confidence in EM. Although many believe that RFID implant tracking will remain science fiction for the foreseeable future – implants cannot (as yet) be tracked by GPS, but can be tracked by localised terrestrial scanning systems – it is probably in the context of reducing participant dependence that they will eventually be considered for use with offenders (see Monmonier 2002: 135; Wood 2007).

CONCLUSION

The location monitoring of offenders has emerged at the intersection of general developments in geolocation, mobile and virtual communication technology, with specific modernising, technomanagerialist, discourses in the world of offender supervision. The option of location monitoring would not have become possible – though it had been *imagined and desired* – without the infrastructure entailed by the former developments. Given the embeddedness of that infrastructure, the increasing normalisation of locatability as a socio-technical practice and the ease with which virtual and geolocation technologies can be customised for crime control purposes, it seems reasonable to infer that the location monitoring of offenders will expand and intensify. It is both derivative and expressive of the emerging 'surveillance of mobilities' – the monitoring of vast global flows of commodities, people and information. Certainly, the feasibility of satellite tracking offenders, worldwide, will increase as a result of upgrades to the American GPS system, the revamping of the Russian equivalent GLONASS, and the completion of a pan-European network of geolocation satellites, Galileo, in 2014.[3] Successful schemes in one country, promoted by the multinational organisations

3 Europe requires the civilian/commercially run Galileo geolocation system in order to free itself from dependence on the American military-owned GPS system, which, in the interests of (US) security, can – and might – be turned off at will. Galileo's signal strength will also be greater than GPS. After some serious political and financial setbacks, the proposed system is once again on track. The movement-monitoring system is aptly named after the Italian astronomer Galileo Galilei (1564–1642), who deduced that motion rather than rest was the natural condition of existence. Thomas Hobbes famously incorporated this insight into his political philosophy in *Leviathan*, describing imprisonment as 'restraint of motion' (Hobbes 1651/1962).

involved, will generate interest and facilitate take-up in others. Such 'policy transfer' will be mediated by local cultural and institutional arrangements (Jones and Newburn 2007), but it is hard to see how traditional humanistic approaches to offender supervision in any country cannot but be affected by the affordances created by new technology.

Over and above the voluntary locatability strategies adopted by citizens, friends and employers, geolocation – whether using satellites or RFID technology – may well loom larger in the everyday lives of ordinary people through controversial road-pricing schemes designed to reduce traffic congestion and air pollution. Significant government investment in such schemes is indicative of the faith being placed in geolocation technology – one day every single car journey might be monitored, recorded and charged for. One cultural consequence of geolocation becoming a commonplace phenomenon – sometimes convenient to citizens, sometimes irksome – may be a political difficulty in presenting the geolocation of lawbreakers as a distinctly or sufficiently punitive form of offender management. In England and Wales, somewhat against official expectations, conventional EM curfews are not universally regarded by the public as effective or tough penalties, and despite all hype to the contrary, satellite tracking could come to be similarly perceived. Neither public scepticism, however, to which politicians might otherwise be expected to defer, nor the absence of decisive evidence of conventional EM's effectiveness (Mair 2006) has significantly impeded its development in Britain, which does suggest that deeper economic and cultural forces, transcending the vicissitudes of mundane political decision making, are sustaining its momentum.

Nonetheless, in 2004, the expansion of satellite tracking in England and Wales was predicated on the eventual stabilisation of the prison population at 80,000, and now that that commitment has been breached, and prison building is once again under way, there may be no *immediate* funding for something as relatively expensive as satellite tracking, least of all in its real-time version. In addition, subsequent government ministers have been less enamoured of it than the Home Secretary who originally championed it. Under NOMS, however, the private sector is still being envisaged as a provider of innovative approaches to offender supervision (Nellis 2006) and it is unlikely that the companies currently involved in EM will abandon interest in tracking's potential. Leaving aside the sociological arguments deployed here to demonstrate why satellite tracking has momentum behind it, indications that it *might* indeed have a future in England come from once unlikely quarters. In the context of articulating 'a Christian approach to punishment' the Catholic Bishops Conference in England and Wales (2004: 92) quickly endorsed the introduction of GPS tracking of sex offenders because 'such systems allow for a much better balance to be struck between concerns about individual human dignity and public safety than is possible in prison'. More recently, Barnardo's (2006), a long-established British childcare charity with a rich and respected humanist tradition, argued that satellite tracking of sex offenders and increased use of lie-detection technology were *preferable alternatives* to the introduction of an American-style (Megan's Law) community notification system. Such support for tracking technology from *within* the Christian/humanist tradition of penal reform tellingly illustrates the extent to which geolocation is nowadays being *imagined* – not just in government or in the commercial sector but in civil society more generally – as something integral to the modernisation of criminal justice and public protection.

Postscript

In May 2008 The Ministry of Justice (successor to the Home Office in England and Wales), announced that it had no plans for a national roll-out of satellite tracking, but added that 'the technology will be kept under review as it develops futher'. Probation officers welcomed the decision, although the decision reflected a considered government decision to switch resources back to prison expansion, away from innovative community penalties. (*Napo News* 199: 2)

References

Aas, K.F. (2007) 'Analysing a world in motion: Global flows meet "criminology of the other" ', *Theoretical Criminology*, 11.

Aas, K.F. (2005) *Sentencing in the Age of Information: From Faust to Mackintosh*, London: Glasshouse Press.

Barnardo's (2006) *A Risk Too High? Would public disclosure (Sarah's Law) protect children from sex offenders?*, London: Barnardo's.

Bennett, C.J. and Regan, P.M. (2004) 'Editorial: Surveillance and mobilities', *Surveillance and Society*, 1(4): 439–445.

Bigo, D. (2006) 'Security, exception, ban and surveillance', in Lyon, D. (ed.) *Theorising Surveillance: The panopticon and beyond*, Cullompton: Willan.

Birkett, S., Ballard, J. and Smith, C. (2007) 'Offender tracking in England and Wales 2004–2006', Paper presented at 5th Conference Permanente Europèene de la Probation Electronic Monitoring Conference, The Netherlands 10–12 May.

Blunkett, D. (2001) *Politics and Progress: Renewing democracy and civil society*, London: Politicos.

Carter, P. (2004) *Managing Offenders, Reducing Crime: A new approach*, London: The Prime Minister's Strategy Group (The Correctional Services Review).

Catholic Bishops Conference in England and Wales (2004) *A Place of Redemption: A Christian approach to punishment and prison*, London: Burns and Oates.

Christie, N. (2000) *Crime Control as Industry: Towards Gulags, western style*, London: Routledge, 3rd edition.

Crawford, A. (2003) ' "Contractual governance" of deviant behaviour', *Journal of Law and Society*, 30(4): 479–505.

Elfers, H. (2004) 'Decision models underlying the journey to crime', in G. Bruinsma, H. Elfers and J. de Keiser (eds) *Punishment, Places and Perpetrators: Developments in criminology and criminal justice research*, Cullompton: Willan.

Elzinga, H.K. and Nijboer, J.A. (2006) 'Court orders, probation supervision through GPS', *European Journal of Criminal Law and Criminal Justice*, 366–388.

G4S (2005) *Satellite Tracking: An introductory guide*, Manchester: G4S.

Gable, R.S. and Gable, R.K. (2007) 'Increasing the effectiveness of electronic monitoring', *Perspectives: The Journal of the American Probation and Parole Association*, 31(1): 25–29.

Haggerty, K. (2006) 'Tear down the walls: On demolishing the panopticon', in D. Lyon (ed.) *Theorising Surveillance: The panopticon and beyond*, Cullompton: Willan.

Hannam, K., Sheller, M. and Urry, J. (2006) 'Editorial: Mobilities, immobilities and moorings', *Mobilities*, 1(1): 1–22.

Henderson, S. et al (2007) *Inventing Adulthoods: A biographical approach to youth transitions*, London: SAGE.

Hobbes, T. (1651/1962) *Leviathan*, New York: Collier Books.

Home Office (2004a) *Reducing Crime, Saving Lives: The government's plans for transforming the management of offenders*, London: Home Office.

Home Office (2004b) *Confident Communities in Secure Britain: The Home Office Strategic Plan 2004–2008*, Cm 6287, London: Home Office.

Jain, J. (2006) 'Bypassing and WAPing: Reconfiguring timetables for "real-time" mobility', in M. Scheller and J. Urry (eds) *Mobile Technologies and the City*, London: Routledge.

Jones, T. and Newburn, T. (2007) *Policy Transfer and Criminal Justice*, Buckingham: Open University Press.

Lianos, M. and Douglas, M. (2000) 'Dangerisation and the end of deviance: The institutional environment', in D. Garland and R. Sparks (eds) *Criminology and Social Theory*, Oxford: Oxford University Press.

Lilly, J.R. and Knepper, P. (1993) 'The corrections-commercial complex', *Crime and Delinquency*, 39: 150–166.

Lyon, D. (2002) 'Editorial: Surveillance Studies: Understanding visibility, mobility and the phenetic fix', *Surveillance and Society*, 1(1): 1–7.

Mainprize, S. (1996) 'Elective affinities in the engineering of social control: The evolution of electronic monitoring', *Electronic Journal of Sociology*.

Mair, G. (2006) 'Electronic monitoring in England and Wales: Evidence-based or not?', *Criminology and Criminal Justice*, 5(3): 257–277.

Miedema, F. and Post, B. (2006) *Evaluation of GPS Monitoring of Offenders in The Netherlands: English translation of the summary and conclusion*, Nijmegen: ITS.

Mitchell, W.J. (1999) *E-topia: Urban life, Jim, but not as we know it*, Cambridge, MA: Massachusetts Institute of Technology.

Molz, J.G. (2006) ' "Watch Us Wander": Mobile surveillance and the surveillance of mobility', *Environment and Planning A*, 38(2): 377–393.

Monmonier, M. (2002) *Spying with Maps: Surveillance technologies and the future of privacy*, Chicago: University of Chicago Press.

Nash, M. (2006) *Public Protection and the Criminal Justice Process*, Oxford: Oxford University Press.

Nellis, M. (2008) 'Electronic monitoring and penal innovation in a telematic society', in J. Doak, P. Knepper and J. Shapland (eds) *Urban Crime Prevention, Surveillance, and Restorative Justice: Effects of social technologies*, Boca Raton, FL: Taylor and Francis.

Nellis, M. (2006) 'NOMS, contestability and the process of technocorrectional innovation', in M. Hough, R. Allen and U. Padel (eds) *Reshaping Probation and Prisons*, Bristol: Policy Press.

Nellis, M. (2005) ' "Out of this World": The advent of the satellite tracking of offenders in England and Wales', *Howard Journal*, 44(2): 125–150.

Nellis, M. (2004) *The Satellite Tracking of Offenders: A brief literature review*, unpublished report.

Nellis, M. (2003) 'Electronic monitoring and the future of probation', in E. Chui and M. Nellis (eds) *Moving Probation Forward*, London: Pearson/Longman.

Nellis, M. (2002) 'Community justice, time and the new National Probation Service', *Howard Journal*, 41(1): 59–86.

Nellis, M. (1991) 'The electronic monitoring of offenders in England and Wales: Recent developments and future prospects', *British Journal of Criminology*, 31(2): 162–185.

NOMS (2005a) *EM Tracking Pilots: Guidance on the piloting of the satellite tracking technology to monitor exclusion orders and prisoners on licence*, London National Offender Management Service, paper no PB7/08.05 version 2:1, draft 10 July 2005.

NOMS (2005b) *EM Tracking Pilots: Service processes*, London National Offender Management Service, paper no PB7/09/05 version 2:1, draft 10 July 2005.

Probation Circular 41 (2004) 'Initial guidance for the prolific and other priority offender strategy: Catch and convict framework', London: National Probation Directorate.

Probation Circular 28 (2003) 'Victim contact work: Guidance on recent court judgements (regarding exclusion zones)', London: National Probation Directorate.

Rengert, G. (2004) 'The journey to crime', in G. Bruinsma, H. Elfers and J. de Keiser (eds) *Punishment, Places and Perpetrators: Developments in criminology and criminal justice research*, Cullompton: Willan.

Rhodes, W.M. and Conly, C. (1981) 'Crime and mobility: An empirical study', in P. Brantingham and P. Brantingham (eds) *Environmental Criminology*, Prospect Heights, IL: Waveland Press.

Scheller, M. (2004) 'Mobile publics: Beyond the network perspective. Environment and Planning D', *Society and Space*, 22: 39–52.

Securicor (2004) *Summary Review of Tracking Use in USA*, Manchester: Securicor.

Shute, S. (2007) *The Satellite Tracking Pilots in England and Wales: Research summary*, London: Ministry of Justice.

Virilio, P. (1995) *Speed and Information: Cyberspace alarm*, CTheory (www.ctheory-net/articles.aspx?id=72).

Weisburd, D. (2004) 'The emergence of crime places in crime prevention', in G. Bruinsma, H. Elfers and J. de Keiser (eds) *Punishment, Places and Perpetrators: Developments in criminology and criminal justice research*, Cullompton: Willan.

Wood, D. (2007) ' "I've got you under my skin": Issues with implants', paper presented at *ESRC e-society Seminar*, London, 12 April.

Chapter 6

Empowered watchers or disempowered workers?

The ambiguities of power within technologies of security

Gavin John Douglas Smith

INTRODUCTION

Whilst the study of surveillance is now a growing global field, there is a distinct scarcity of empirical and theoretical attention focused on the activities and roles of those actually operating the various technologies of security in existence. This is a critical omission, as the everyday actions and behaviours of such individuals significantly affect the overall *practice of surveillance*, a key substantive topic assumed in much theory. Research which has been done generally follows, intentionally or inadvertently, a Foucauldian tradition emphasising the *power*[1] of the watchers over the subjects of their gaze. Watchers of closed-circuit television (CCTV), for example, are said to be 'empowered' through their apparent God-like ascriptions of unhindered, anonymous, unilateral vision and asymmetrical informational knowledge about certain populations (i.e. capacity to know). They also accrue power through their subjectivised jurisdiction over the behaviour of citizens (i.e. capacity to define a situation), their positional autonomy and influence over the technological systems through which they gaze (i.e. capacity to control), and their oversight of civil order in public space (i.e. capacity to make a transformational difference).

Yet this depiction of operators as empowered, disciplinary agents rests upon a one-dimensional view of their role in surveillance systems as simply *watchers*. Whilst such individuals are unquestionably ascribed with unique observational capabilities and a particularly distinctive authoritarian position, detailed ethnographic research with CCTV operators reveals that

1 Classically one of the most contested concepts in the social sciences, 'power' or 'empowerment' in this chapter refers to all or any one of the following capabilities (note that the first and last definitions presented here emanate from the ideas of Foucault (1972; 1980) and Giddens (1984)): the capacity of an actor *to know*; the capacity of an actor *to define a situation*; the capacity of an actor *to authoritatively influence or exert control*; and the transformational capacity of an actor to intervene in a given set of events *to achieve outcomes or to make a difference*.

their role in such systems of control is characterised by much greater complexity. They are watchers, but also *workers*, subjected not only to the same capitalist regimes of domination as any other labourer in late modernity but also to an emotive duress produced by the very technologies which earn them their living. Disempowerment is further compounded both by the operators' structural and hierarchical impotence over other agents of control within the 'surveillance web' (McCahill 2002) and, amending Giddens's (1984: 14) classic definition of agency, through their *in*capacity to make a *physical* difference in the mediated action taking place on the screens. Yet CCTV operators are not simply mindless robots or automated machines – they are of course *humans*, equipped with enough spirit and creativity to, in an 'unofficial role', ethnomethodologically reassert and project, through interaction, their emotional subjectivities and desires over the technologies they operate in a subjectively meaningful, but organisationally contradictory, fashion (Smith 2007a).

This chapter argues, therefore, that surveillance operators should be seen as simultaneously empowered, disempowered and re-empowered by the conglomeration of technologies they operate and through the myriad mutating roles they arbitrarily perform. As will be shown, power within such workplace settings is a fluid, indefinite and ambiguous property. Indeed, in a substantive area too often dominated by abstract discussions emphasising the technological, the automatic and the futuristic, the key aim of the author throughout is to return the *social*, the *empirical* and the *multifarious* to the forefront of surveillance analysis (Latour 2005).

In order to arrive at this position, however, the reader will initially be taken through the central arguments which collectively situate the CCTV operator as empowered agent of control. Whilst insightful, valuable and indubitably valid to some degree, a focus only on the 'official role' of CCTV operators as *watchers* means that such claims have a tendency to overemphasise the power which such individuals actually possess. Nonetheless, the discussion usefully brings out the rationale for this chapter, namely, to investigate ethnographically the extent to which public space operators themselves experience empowerment through their Panoptic placement within the 'room of control' (Smith 2007a: 281). Attention turns, in the findings section, to a two-pronged ontological re-conceptualisation of the CCTV operator, by empirically taking into account the latter's other roles in the setting as *worker*, and the various regimes of systemic control, emotional and 'structurational' disempowerment that this identity implies, and *human agent*, equipped with the creative dynamism to unofficially attain re-empowerment through interaction with the system. The chapter concludes with a short summary of the main issues and their various implications for surveillance studies.

CCTV OPERATORS AS EMPOWERED AGENTS OF CONTROL

Being a public-space CCTV operator is a job like no other. It often involves the control of, in a directorial sense, large networks of cameras which compress time and space (Giddens 1990). As such, camera operators gaze upon multiple realities, accumulating in the process vast information, knowledge and intelligence from the objects, bodies and behaviours viewed. They find themselves locked within a continual interpretive figuration of risk assessment and management (Beck 1992).

Indeed, operators can usefully be conceptualised as *artists*, whose role it is to both detect and respond to any 'blots' on a social canvas composed of consumerist, aesthetic and civil-order ideals. In this sense, CCTV staff are symbolically ascribed the *visual* guardianship of the urban social order, a potentially enthralling but pressured position. Through the medium of largely wired (closed) but increasingly 'wireless' (open) camera networks, such actors, housed in remote monitoring 'towers' safely distanced from the action, are presented with the unique opportunity to freely, discreetly and subjectively select, target and scrutinise all that crosses their gaze. From this privileged position, everyone and everything becomes visible, but at the same time, a potential risk (Neyland 2006b). Advanced communicative technologies facilitate the operators' direct contact with embodied agents, or 'corporeal guardians', of control on the ground, e.g. police, security staff etc., effectively enabling them to alert the latter individuals to any action or behaviours deemed obstructive or threatening to the flow of consumer-driven day- and night-time economies. Perhaps it is unsurprising that many scholars ascribe actual and symbolic power to the role of CCTV operator. Indeed, four main strands in the pertinent literature, each of which will be reviewed below, give considerable credence to such a position – the CCTV operators' ability to see and know, their subjective capacity to interpretively define and edit the images and realities watched, their pivotal role as 'risk-assessing' informants and gatekeepers to emergency-service resources, and their aptitude to undermine the operational power of surveillance.

Vision and knowledge

Norris and Armstrong (1999: 150) argue that the form of 'one-way' monitoring enjoyed by CCTV staff affords such watchers a *legitimated* voyeuristic power unmatched by any previous urban authority or embodied citizen:

> The difference between the unmediated gaze of the eyes, and the camera mediated gaze of the CCTV operative, [lies in] the profound asymmetry of power inherent in CCTV monitoring . . . [in that] the veil

> of the camera denies the possibility of a reciprocal exchange of visual data.
>
> (Norris and Armstrong 1998: 5)

For these authors, the introduction of CCTV cameras to city-centre streets has fundamentally altered the nature of urban micro-sociological relations between state and citizenry, from an authoritarian gaze which was historically embodied and face-to-face to one which is now 'distanciated' (i.e. physically removed in space), disembodied, anonymous and technologically mediated (Giddens 1990; Dubbeld 2003).

Indeed, the operators' placement behind the screens enables such individuals to 'virtually' follow, track and facilitate the exclusion of any bodies perceived to be 'out of place or out of time'. CCTV observation, then, it is argued, precludes the possibility of the person observed directly questioning or physically challenging the watcher and this very process produces an uneven, unilateral power imbalance between the two (Dubbeld 2003). So whilst camera operators can apparently freely choose to monitor in close detail who and what they like, they themselves cannot be seen, remain relatively unaccountable for their gaze and are, in a corporeal sense, invisible.

It is not only vision which is said to empower the watchers but also their capacity to collect, analyse, store and *know* intimate details about the populations they observe (Foucault 1977; Fyfe and Bannister 1996; Dubbeld 2003). Operators often have access to large security databases containing sensitive and detailed information on many socially ascribed deviant groups. Such electronic profiles are regularly updated and typically include personal details of where the 'Stars of CCTV'[2] reside, work, frequent and operate, when they were born and into which family, who they associate with and where they were last seen. The records further contain an individual's criminal history and any 'markers' for relevant health, behavioural and social welfare conditions. My own research showed that it was routine for operators in a variety of the settings visited to possess official and/or unofficial dossiers ('rogues galleries') containing the pictures (often digital stills taken from the CCTV cameras) and addresses of certain persons, with abstract commentaries detailing why such individuals were of interest (see also McCahill 2002). Indeed, some operators were even permitted to conduct their own Police National Computer (PNC) checks on 'Stars' they had located, either to establish whether there was an outstanding arrest warrant issued or to update

2 Smith (2007a) coined the term 'Stars of CCTV' to refer to the individuals and groups who are extensively watched by the operators for either their criminal activity or 'idiosyncratic' behavioural antics.

current files (e.g. 'so and so is out of prison now'). As one operator informed me:[3]

> Yeah I've learnt most of their [the Stars of CCTV] dates of birth off by heart [through having to habitually check to see if they are 'wanted' on the PNC]. Pretty sad, isn't it?! It's funny, I know more about the criminals than I do about some of my family.
>
> (Operator 5)

Unrestricted access to both constant police dialogue over the radio and the day- and night-time private security firms' regular informational interchanges means that not only can operators freely inspect people in detail, they can also know a great deal about some of them. For this reason, theorists have likened CCTV operators to the all-seeing, all-knowing prison guards anonymously located in Bentham's Panopticon tower of rationalised control (Fyfe and Bannister 1996).

Defining and editing realities: subjective power

Numerous cross-cultural studies have been conducted focusing attention on who and what is 'surveilled' by CCTV operators in differing organisational settings and, crucially, which interpretive schemes or 'working rules' guide operator decision making (Norris and Armstrong 1999; McCahill 2002; McCahill and Norris 2003; Lomell 2004; Smith 2004; Neyland 2006b). Such research has shown conclusively that suspicion is 'socially constructed', with CCTV operators predominantly *choosing*, as a result of their subjectivities and the workplace culture in which they are embedded, to target and associate criminality with young, working-class males, ethnic-minority populations, sub-cultural groups and particular forms of immobility:

> The gaze of the cameras does not fall equally on all users of the street but on those who are stereotypically predefined as potentially deviant, or who, through appearance and demeanour, are singled out by operators as unrespectable. In this way youth, particularly those already socially and economically marginal, may be subject to even greater levels of authoritative intervention and official stigmatisation, and rather than contributing to social justice through the reduction of victimisation, CCTV will merely become a tool of injustice through the amplification of differential and discriminatory policing.
>
> (Norris and Armstrong 1997: 8)

3 For reader clarity, the ethnographic data is presented in the following way: paraphrased commentaries and field note extracts appear in *italics*; direct quotes from CCTV operators and managers in the form of written statements or taped interviews are portrayed in regular font style. All excerpts are indented.

Smith (2007a: 300) has brought attention to the discretionary 'definitional' powers which CCTV operators possess regarding what constitutes, in their eyes, 'real crime'. He suggests that such individuals are not simply passive conduits of information, showing instead how their personal judgements can actually influence, sometimes determine, the outcome of surveillance and the course of justice administered on the streets. For Smith, CCTV operators are *active* agents of social construction who, like television audiences, partake in extensive practices of phenomenological hermeneutics. He brings out this point by empirically detailing the ways in which such actors interpretatively create biographies, narratives and shared identities for those watched, enabling anonymous bodies in space to become personalised, individualised and subjects of interaction. Such story telling, it is argued, also enables meaning and entertainment to be gleaned from the screens, and behaviours to be contextualised and thus better conjecturally understood.

When critically investigating claims regarding their authoritative empowerment, the function of CCTV operators, not only in actively constructing but also in 'editing' reality, should not be overlooked. The moment a criminal or anti-social incident occurs in view of a camera system, it is the watchers who are entrusted, through their organisationally ascribed professional role as both semiotic 'experts' and pictorial 'custodians', to use a range of tacit ethnomethodological skills (interpretative, detective and technical) to fuse together pieces of reality (evidence) in a sequential montage (production). This involves initially locating footage disconnected in time and space, artificially and meticulously reassembling it, the consequence of which is a bricolage of temporally and spatially distanciated realities, a socially induced transformation of pictorial meaning and a synthetic set of simulated actions which help tell a particular 'story' (Neyland 2004; 2006a). Despite this practice being open to considerable interpretation, manipulation and abuse, legal prosecutors use such images in the criminal courts as an 'expert witness' account or as realist evidence depicting objective truth, overlooking the fact that it has been produced through the subjective judgements and decision-making processes of socially biased CCTV operators. As editing shapes the voting patterns of the viewing public in reality-television programmes such as *Big Brother* (Andrejevic 2004; Hill 2005; Biressi and Nunn 2005), it would not be unreasonable to assume that an operator-constructed depiction of events shown in court must also directly influence the views and 'voting' processes of judges and jurors in criminal cases of the highest severity. Thus, from distanciated rooms of control, and in their official role as *watchers*, CCTV operators not only possess the agency to subjectively construct suspicion and narratives for the realities watched, they also have the genuine capacity to influence, through visual editing, the outcomes of both social *and* criminal justice.

Clearly, findings such as those above emphasise the interpretative *power* of the watchers in both the construction of reality and decision-making processes. Operators, however, are not only legitimated 'reality interpreters' with the capacity to edit, they are also interpretivist 'risk assessors'.

Expert risk assessors: informants and gatekeepers

CCTV operators, though largely untrained intellectually in the physical and psychological dynamics of human behaviour, by way of their position as watchers officially perform the job of risk assessing 'informants'. They are the ultimate 'super narks', drawing upon their tacit and experientially built-up evaluative skills, garnered from prolonged viewing of situations and reality, to inform the police about individuals or situations they deem to be threatening to the consumerist/civil order. This often involves CCTV staff contacting police control and alerting them to an incident they are watching which, they feel, requires a police presence urgently.[4] In turn, police control regularly asks operators to monitor confrontational, precarious or suspicious situations and report back should things escalate. Not only do operators notify and inform, they also regularly take on a 'logistical' role, directing in real time both grounded and mobile officers to the location of assailants, victims and speeding cars. Another facet of the informant role empowering operators occurs when police officers contact the former asking whether an incident was captured on camera. If the event has been witnessed, more often than not (certainly in my research) it is the operator's information which directly determines the outcome, e.g. an individual's arrest.

CCTV operators often act as police resource 'gatekeepers' to the private agents of control operating within the surveillance web. Thus, should a security guard or door steward require a police unit to attend an incident, they must request it from the CCTV staff by radio. However, an over-supply of requests, a limited number of available police units and a convoluted communicational chain of command mean that, in practice, operators, from the safety of the room of control, form their own personalised risk-assessment judgements regarding whether or not the call merits a police response. Perhaps unsurprisingly, the result of such distanciated and subjective decision making, based as much on structural location, workplace cultural misunderstanding and historical relations as on accurate interpretation of real risk, is often an 'unavailable' police presence, a situation which serves only to generate further interpersonal tensions, distrust and disillusionment between the various parties (Smith 2007b).

Such cultural tensions are also indicative of CCTV operators' ability as watchers to cause dissensions *within* the systems they inhabit.

4 The operators have no directional power to actually *order* the attendance of any emergency service.

Workplace conflict and organisational subversion

A fourth area of investigation theorising the agency and capacity of CCTV operators focuses on the ability of such watchers to deliberately neglect or subvert surveillance systems through a process called the 'human mediation of technology' (McCahill and Norris 2003: 46). Many empirical studies highlight the importance of the workplace culture and wider social relations in determining the daily interactional outcome of CCTV operation, particularly drawing attention to how the extent of strained work relationships and negative structural ergonomics can drastically undermine the organisational goals, effectiveness and cohesion of surveillance systems (McCahill and Norris 2003; Smith 2004; 2007a; 2007b). McCahill and Norris's (2003) study of a south London shopping mall, for example, demonstrated that the persistence of organisational conflict between security officers and middle management, and between security officers and the local police force, had created a divided and disillusioned workforce, resulting in a cultural dynamic of laziness, frustration and discord within the locale. When combined with the differing personal characteristics and goal orientations of each security guard, the culmination was general 'non-use' of the CCTV system. Norris and Armstrong (1999) have also found distrust and poor institutional integration between local authority-employed operators and the police, a crucial determinant factor behind low CCTV prosecution rates and restricted police deployment to operator-located incidents.

A further social process affecting CCTV operation can be seen in McCahill's (2002) research, which revealed how management's use of surveillance technology to monitor and control employees was undermined and resisted by security staff and lower-level workers due to their shared class identity. Security staff, culturally, had more in common with and greater sympathy for their equivalently paid colleagues and their informal practices than they had with the goals and actuarial controlling strategies implemented by a top-heavy management. This led to them deliberately ignoring the informal, often illicit, workplace activities in which the workers were involved.

Clearly in these cases, the operators are empowered in their capacity to 'make a difference' in terms of refraining from operating the system or sabotaging its intended functions.

Missing pieces

The above examples illustrate the empowerment of CCTV operators through their official 'role' as watchers, drawing attention also to the plethora of extended opportunities afforded such a position. Yet operators are not simply watchers, they are also workers and human beings. They are both paid labourers constrained by their socio-spatial environment and the organisational rules and regulations of the milieu, and reflexive human agents, in other words, emotionally embodied citizens with distinctive goals, drives and capabilities. CCTV rooms are not simply voyeuristic paradises, rather they are workplace settings subject to the same unrealistic organisational goals and demands, forms of exploitation and regimes of domination as any other. In this light, the role of controlled and repressed worker/employee quickly ousts the untamed and unbridled freedom of 'watcher' alluded to in much of the discourse above. Moreover, watching is not always necessarily empowering, it can be difficult and traumatic work.

Yet, by the same token, human agency should not be underestimated, and such locales also produce conditions and opportunities for knowledgeable, creative and skilful social actors to flourish. Perhaps ironically, re-empowerment can be found *through* the very technology which in part creates operational disempowerment. Indeed, far from being a unilateral and static property, power in this milieu, much like a restless tide, flows back and forth in interaction between subjectivity and screen, self and spirit, body and bureaucratic organisation. Power's dialectic nature leaves the outcome of such relations in constant and indeterminate flux (Giddens 1984).

CONTROLLED WORKERS AND EXPERIENCES OF POWERLESSNESS

The central rationale for writing this chapter came from a contradiction which arose in the early stages of my research on CCTV operation. Having read some key texts on the theory and practice of public surveillance, I had entered into the field expecting to find a collectivity of instrumental, almost automated, *watchers* equipped with great powers of vision and knowledge. I was actually confronted, however, with a series of narratives, accounts and behaviours from *workers*, evidently disempowered by, frustrated with and alienated from their labour. Such powerlessness related to four principal factors: the wider ontology of control underpinning CCTV surveillance which constrains operational freedom (domination), the operators' position at the bottom of a social control hierarchy (structure), the operators' inability to take part and intervene in the physical action being watched (agency) and the degree to which their anonymity and integrity is being eroded by escalating court appearances (fear). Each of these issues will now be considered.

External control(s) – domination

CCTV is inextricably connected to its own (and a wider societal) ontology of control, which in turn is linked to power, governance and risk management. In public space, the technology's rapid deployment is tied to strategic spatial-management programmes and political-economic policies primarily concerned with securing predictability and controlling people and movement. Yet the controlling dynamics underpinning CCTV do not start and end simply with the technology and its external electronic gaze. Perhaps ironically, the operators are themselves increasingly subject to a range of *internal* regulations and forms of monitoring within their respective places of work. Such measures include having to adhere to a raft of ambiguous yet legally stringent rules and procedures regarding what is and is not permitted, an expectation of continuous intelligence-gathering productivity and the requirement to meet targets and provide documentary evidence that sufficient intelligence sightings of certain populations have been achieved, and being physically constrained at a 'control' desk in a dark room for long periods of time and for relatively low psychological, financial and organisational reward. Indeed, following a classical Marxian line of thought, such labourers have neither ownership rights over the means of production (the technologies) nor control over the end product of their labour (the images/realities). Perhaps unsurprisingly, then, many experience a distinctive form of alienation emanating from the CCTV cameras, controls and televisual screens themselves and, crucially, the mediated realities they so graphically display.

Never more so is their subjugated role as 'watching workers' brought out than when operators are forced to view particularly distressing footage:

> I've had to review footage recently of two poor souls committing suicide. I didn't want to, but it was my turn. It's really not a nice job, but the police need it done to check there are no suspicious circumstances. You've just got to try and not let it get to you, though it's hard.
>
> (Operator 20)

CCTV operators, in other words, have little agency to 'avert their gaze'. They must film and observe everything despite the severity or banality of the scene, and are thus effectively controlled by the screens. The job of watcher implies just that: compelled watching of the variable and the mundane, the shocking and the prosaic. Perhaps it is the stark contrast provided, and an addictive fascination with the unpredictable, which sustains their interest? Whatever the reason, it is the obligated nature of the gaze and its exposure to scenes of graphic incivility which are crucial to understanding why such surveillance workers need to be seen as performing, in their everyday lives, a distinctive form of emotional labour, work and management (Hochschild 1979; 1983).

Indeed, the general assumption in the literature that operators are free to observe who and what they choose is misleading in other ways. Operators in my research were often *told* who or what to watch, and even challenged by superiors[5] wishing to know why they were monitoring particular images. The following field extract was a fairly typical example:[6]

5 Superiors in the various systems researched included police controllers (i.e. those who despatch and direct police units), police officers, council supervisors and management.

6 Superiors, by way of a router link in the system, could also view the camera images being produced on a separate screen in their respective locations.

> Operator 11 receives a radio call from a security guard alerting him to a 'dodgy' suspect. The operator follows the male in question for around five minutes. A short time later, Operator 10 receives a call from the police controller asking why Operator 11 was watching the man ... Operator 11 is clearly angered by this call and asserts, 'There's too much accountability with this job.'

Hence it appears that CCTV operators in practice, at least to some extent, are answerable for the gazes they cast.

It is not simply the setting, technology and organisational rules which constrain operators; their structural position in the wider surveillance web also reinforces feelings of powerlessness.

Organisational impotence – structure

CCTV operators, in actuality, form a relatively small part of a much wider urban surveillance web, a surveillant assemblage composed of many different agents of social control from the public and private sectors, e.g. police officers, private security staff, etc. (McCahill 2002). Whilst being, operationally, at the interactional nucleus of the web as watchers, such workers occupy, hierarchically, the lowest position in a structured social control chain of command. Indeed, officially and organisationally, CCTV staff have neither directional nor authoritative power over *any* of the other embodied actors they work alongside. Moreover, operators also find themselves to be the constant subjects of other people's ordering and control (often the very people they gaze upon!), being regularly and brusquely instructed what to do and where to look:

> [Police officer to the operators] 'Right, just to make you aware, from now on you're not to touch cameras 4, 15, 38 and 51 as they're set on ANPR [Automatic Number Plate Recognition] duty, ok?'
>
> [Doorman to the operators] 'CCTV, put your cameras on our front doors, we've got a male kicking off.'

Despite unrivalled power of vision, the structural placing of operators at the very bottom of the command hierarchy creates rooms of control which are suffused with strong negative emotions, particularly feelings of powerlessness, frustration and resentment. The following extract is evidence of this:

> Two police officers enter the CCTV room and begin verbally instructing the operators where to focus the cameras. When the two constables leave, Operator 1 blurts out, 'How dare they do that – I hate when the

police interfere with our work. They think they can just come in here and take control of everything. They've got no right to do that.' Operator 16 nods his head, stating: 'It used to be a lot worse. I remember times when they [i.e. the police] actually pushed you aside and took control of the cameras . . . It's one of the reasons why I can't be doing with this job any more.'

Indeed, intense feelings of irritation associated with directional disempowerment are further generated when operators, in the hope of having a unit deployed, alert police control to an incident but regularly have their concerns either ignored or simply dismissed:

> Operator 8 contacts police control informing them that there are 'four guys squaring up and a unit down there sharpish should stop things escalating'. The operator is told to 'keep watching it' and to let control know if things progress. The operator slams down the radio in disgust: 'What a waste of time, what's the point? You can see how much they value my judgement!'

As operators are ascribed the paradoxical role of police resource 'gatekeepers' without having any actual authority over the deployment or ordering of units, they are regularly forced into a difficult mediating position, with door staff on the ground impatiently requesting updates on units which the police controller has either refused to despatch or simply does not have (Smith 2007b). Thus, while operators are empowered through their ability to *socially construct realities*, they are simultaneously disempowered through their inability to *systemically instruct resources*.

CCTV operators are also powerless in respect of any managerial change occurring within the rooms of control they inhabit:

> Operator 6 informs me that the operators have a specific clause in their contract under the heading, 'Any Other Duties', meaning management has the power to introduce new equipment (i.e. more cameras and radios, etc.) into the CCTV suite without having to increase the operators' salaries for any extra work which this would involve: 'That basically means they can put in more cameras, but they don't have to give us extra pay to monitor them. It's not right.'

Likewise, echoing Operator 6's comments, they are compelled to perform a plethora of practices which were never part of their original job description or indeed training:

> Due to the police controllers being short staffed, Operator 2 has been asked to help traffic division conduct PNC vehicle registration checks on

> the cars stopped at their checkpoint: 'We shouldn't have to do vehicle checks as that's not our job. We were never trained to do half the things we actually do around here. We always have to do extra tasks 'cos they're short staffed.'

It is not only an organisational impotence that operators experience; such workers also feel, contrary to the literature, a deep sense of powerlessness precisely through being distanciated and physically separated from the action unfolding.

Physical impotence – agency

Not only do CCTV operators lack directional governance over other agents in the surveillance web, they have no control or power over the social action taking place on the screens. While they can indirectly influence events by alerting the pertinent authorities, they are compelled to passively watch and record a variety of unpredictable scenes and realities in which they can neither directly intervene nor make a meaningful physical difference. Indeed, they rely and are entirely dependent on *others* to materially act on their behalf, others who are often oblivious to the chronology, speed or seriousness of the incidents unfolding. Unsurprisingly, this can be a particularly traumatic and frustrating position to occupy:

> Some of the stuff you have to see can be quite tough to watch, especially since it's happening live.
>
> (Operator 13)

> Yeah, I agree. You know, it's not nice watching a guy getting his head kicked in while you're waiting for the unit to arrive. You can't do anything and it can be pretty frustrating and distressing at times. At least when you're reviewing tapes you kinda know what to expect, you've got pre-warning; but when you're watching the incident unfolding in real time, it's a different story. I mean, I've sat here before and watched a guy jumping [i.e. committing suicide] in front of my very eyes on the camera. That was pretty hard to take. It's really difficult because you know you can't do anything, but yet you have to watch. It's a horrible feeling.
>
> (Operator 9)

As these narratives suggest, being mere passive observers of mediated space, CCTV operators are continually experiencing a form of role conflict in their working lives, where an official remit consisting of active 'public protection' and 'crime prevention' is shattered by the realities of their physical impotence. Indeed, institutional pressures to capture the extreme, and the cameras' facilitation of graphical intimacy, mean that operators are periodically subjected to situations of the most atrocious human suffering, about which they can do nothing.[7] Their gaze habitually finds individuals requiring care, compassion and physical safeguarding which is beyond their means. Thus, despite being introduced as social deities to actively rescue vulnerable urban citizens, operators are in fact the helpless and immobilised watchers of society's distress. The key elements of power, as defined earlier, seem a world away from this particular position of paralysis. Unsurprisingly, those choosing to disregard the mediated barrier between themselves and the action find the actualities of the role especially difficult to fathom:

> You just feel sometimes that you want to help out but all you can do is sit back and watch until the police arrive. It can seem like such a long time, especially if someone is getting jumped on or is threatening suicide. You just feel powerless and a bit guilty that you can't do more.
>
> (Operator 7)

The key point is that the very distanciation which facilitates the role of operators as empowered watchers simultaneously creates in the workers feelings of profound powerlessness.

Some operators, however, manage to negotiate this positional anomaly by psychologically implicating themselves, almost as quasi-participants, in the action unfolding, deriving a sense of phenomenological satisfaction that they have made a 'valuable difference'. This is evident in the ethnomethodological narratives operators tell, particularly in the *active* and *involved* language they employ when describing their vocational 'heroics', indicative of an objectively fallacious, but subjectively meaningful, belief in the extent of the 'physical' part actually played in the action:

> Oh you really should have been here last night. We got two drunk drivers stopped and arrested and got a guy in custody for a robbery; we've had a really good week.
>
> (Operator 3)

Interestingly, the growth in 'Talking CCTV'[8] systems is welcomed by many operators, who feel that it will provide an extra 'physicality' in events:

> Yeah, I think it would be good if we had them. I'd love to say, 'Oi, stop that, you' and just tell some of those idiots to 'wise up'. I'd prefer

7 Indeed, as an 'absent presence' the operators are removed corporeally from the action, but are intimately present in a mediated sense.

8 Systems fitted with audio facilities which enable operators to both listen to and talk directly with those they are observing. See http://news.bbc.co.uk/1/hi/england/6524495.stm (accessed 10 April 2007).

> cameras with a laser-gun facility on the top, though! That would be much better.
>
> (Operator 1)

Having the ability to verbally communicate with the watched is, for the operators, less about the official crime-prevention rhetoric espoused by government and managers than about the breaking down of felt distance between themselves and the action. Such a mechanism would, in effect, further facilitate the norms of more routine and meaningful face-to-face interaction (Goffman 1967; 1971), particularly for those feeling the psychological strains of simulated disembodiment and physical impotence. It would appear, then, that the operators themselves are dissatisfied with being mere watchers, transcending the limitations of such a role through a multiplicity of differing phenomenological measures.

Going to court – fear

CCTV operators are now being asked to attend the criminal courts more regularly to provide judge and jury with 'expert witness' accounts regarding captured footage.[9] Most of the operators spoken to feared going to court, finding such an event stressful, intimidating and disempowering. This was for two main reasons. First, by giving evidence, operators' identities become publicly known, eroding the 'anonymous watcher' privilege so crucial to the augmentation of empowerment. Indeed, some have to give evidence relating to the filming of serious crimes, others to the wrongdoing of those who reside in nearby estates. In many cases, the combination of an operator's footage and narrative is enough to convict and send individuals to prison for lengthy sentences. For these reasons, several operators reported understandable feelings of vulnerability:

> The thing that worries me about it [i.e. going to court] is if I was giving evidence and the accused recognises me later. I mean, I've managed to get a guy prosecuted who lives near me; he frequently speaks to me when I bump into him in my local shop, having no idea who I am, or that I've had him lifted!
>
> (Operator 4)

> I had to describe the incident to the jury before formally identifying the accused in the dock. He was this big guy and he just sat there scowling

9 This is interesting for theoretical reasons as it perhaps suggests that the objective power of the visual has declined in a society now saturated in mass-mediated imagery, much of which is the subject of editing and manipulation.

> at me. The guy ended up getting put away for a few months, and I always worry when I'm out and about that I'll bump into him, and he'll remember me.
>
> (Operator 2)

> I'm just scared that I'm going to be called to court some time to give evidence and someone's going to recognise me at a later stage. The whole point of being a CCTV operator is that our identities are anonymous and we are hidden from the public. If our faces become publicly known, then our personal safety could be at risk.
>
> (Operator 17)

Second, going to court also involves such workers overtly answering questions relating to the 'production' they have created, often from linguistically skilled and eloquent prosecution and defence lawyers. These can range from the broad (i.e. what happened in the incident) to the more specific and difficult (i.e. why, in particular, the operator decided to film the accused). Despite operators recording the incident in question and producing the ocular evidence, they have often not viewed the footage for over a year and have forgotten the particularities of the case. Defence teams, however, are quick to plug inevitable gaps found in the operators' descriptions, leaving the latter to feel that their integrity and veracity are being directly challenged and called into doubt. This creates feelings of anger and resentment, often prompting a related loss of self-esteem:

> The thing is that we're doing a public service and they treat you like that [poorly]. You feel like you're the one on trial. It's a joke. We shouldn't have to go to court in the first place. I mean, surely the footage is enough and speaks for itself?
>
> (Operator 14)

> There's far more pressure on us now to capture the whole event without missing anything. They [the defence] asked me questions about myself, the job, the actual incident and the production, trying to insinuate that I had made a mistake or that the production was in some way non-credible. Not pleasant.
>
> (Operator 4)

THE HUMAN SPIRIT AND RE-EMPOWERMENT: 'TEMPORARY ESCAPES'

Thus far we have seen the way in which control is the bedrock of surveillance systems, and how operators experience, through their *official* roles as watchers and workers, a duality of both empowerment and disempowerment (often simultaneously). Yet operators are not simply inactive, unthinking organisational robots/dupes; rather their composition as reflexive, emotive and creative social agents, equipped with the capacity and interactive powers to interpret and utilise material resources for particular ends, leads to the adoption of a range of cathartic behaviours as objectively inefficient as they are subjectively understandable. Such re-empowerment takes many forms, ranging from conscious systemic conflict to the more subtle, almost unconscious, manipulations of the system's fundamental properties. The latter practices can be seen as meaningful 'escape attempts' (Cohen and Taylor 1992), albeit temporary in disposition, where the operators interactively, and *unofficially*, reassert themselves over the system by using the technology for their own gratification. This is part of what Smith (2007a: 294) terms 'the humanisation of technology process', that is, the interactive fusion of subjectivities with technologies. The remainder of the chapter empirically considers the forms which such resistance takes.

Humanising the screens: secondary adjustments

The dialectical nature of power allows for subversive capacity (Giddens 1984), Goffman (1961: 171) classically introducing the term 'secondary adjustments' to conceptualise the innumerable (unofficial) ways in which social actors, for a variety of subjective reasons, employ unauthorised means to resist, negotiate or subvert their organisationally defined and attributed primary (official) roles or social identities (for unauthorised ends). CCTV operators re-empower themselves through regular secondary adjustments, using the cameras informally as vicarious agents for resistance, escapism and the projection of fantasy in an organisational setting dominated by control and feelings of powerlessness. Indeed, the human spirit's creative capability for ingenuity and inventiveness and dual detestation of domination and monotony – two key factors underpinning CCTV operation – enable such individuals to discover and locate, through interaction with the technology, a plethora of latent opportunities to re-establish authority, amusement, satisfaction and meaningfulness. While Smith (2007a) has previously shown how the cameras are used as media for virtual communication and interaction, they are also re-appropriated for *aesthetic* and *relaxation* reasons, so that their operators can achieve temporary escape from either the repetitiveness and tedium of a quiet shift or the stresses and pressures associated with viewing a traumatic incident:

> Having been involved in watching several violent assaults during the shift, Operator 1 moves the camera to focus on the river. She admires how still and calm the water is, asking Operator 16 to have a look at this scene on his monitor screen: 'I love it when it's like that, when the water's just completely still. There is not a breath of wind out there tonight.'

> Operator 12 is using the beach cameras, and zooms one of them out to sea: 'What a pretty sky . . . That's one of the small consolations of this job; you get to see all the sunrises over the sea, it's very relaxing. I do like watching the sun come up.'

An appreciative aesthetical 'view' of nature and the natural order, however, can quickly metamorphose into a voyeuristic gaze fixed on the assortment of visual pleasures offered by the objectified human body:

> As Operator 14 is turning the camera, he focuses on the public walkway and beach area and says: 'Ah, this is great along here in the summer, lots of bikinis!'
>
> He's a bit of a looker, isn't he?! I wouldn't say no.
>
> (Operator 1)

Similarly, the operators in McCahill and Norris's (2003: 25) research had a monitor screen removed as 'some of the guards were being a bit naughty and misusing the cameras . . . watching things they were not supposed to be watching: girls walking down the street and things like that'. Indeed, it is clear that, from the physical and symbolic safety of the room, personal gratification and entertainment are acquired from using the cameras to view the graphic, the repugnant and the intimate:

> Operator 21 informs me that he has made up a 'best of fights' tape, of which he is clearly very proud: 'Yeah these are some of my best captures. Pretty good stuff.'
>
> Operator 3 tells me about a male she saw defecating on the street: 'We initially thought he was planning to use the branch he was carrying as a weapon so we kept our eye on him, but it turned out that all he wanted was the leaves to wipe his arse on! It was disgusting.'
>
> 'We saw them having sex just behind there. Yes, we see it all in here, it keeps us entertained!'
>
> (Operator 17)

Albrechtslund and Dubbeld (2005: 220) and McGrath (2004) address the ways in which surveillance technologies can be harnessed as forms of entertainment and how the operators of such mediums can derive significant pleasure from them. The 'fun side of surveillance' is clearly brought out in the following examples of 'surveillance games', played amongst the operators themselves and with those whom they watch:

> A short time into the shift, Operator 14 tells me: 'We've got all sorts of games we play in here. Our current favourite is "identify the building roof".

> 'I much prefer dayshift as there is generally much more going on even if it maybe isn't so explosive. I mean I quite enjoy trying to locate and follow the shoplifters and all the games we play with them. It's quite good fun.'
>
> (Operator 4)

Operators, through secondary adjustments, find many inadvertent 'social learning' utilities for the cameras, allowing them to acquire tacit knowledge, protect their own property, genuinely 'window shop' and even keep updated on sporting events:

> Operator 6 is watching on his monitor screen live coverage of a Premiership football fixture. He has artfully focused a camera on a large plasma screen showing the match above the front entrance to a city pub. He continues to watch, and commentate on the game uninterrupted for the next 25 minutes until its conclusion.

There is a further form of secondary adjustment in which operators partake, one which is both socially and systemically integrative, i.e. functional for both operator *and* organisation (Lockwood 1964). This relates to operators, particularly those who are disabled, utilising the cameras' ability to transcend time and space barriers, thus fostering a mobility not possessed in their personal lives:

> It's like it has given me this great opportunity of moving freely round the city, something I can't really do normally because of my bad legs.
>
> (Operator 20)

In this sense, operators garner adrenalin and excitement from participation in the action taking place using a method unavailable to conventionally embodied actors:

> I love car chases in the city, you have to move from camera to camera second guessing which road he might come out onto. It's like a game of chess or cat and mouse but at speed. I can be anywhere in the city at the touch of a button. It's great.
>
> (Operator 19)

As the above quotes demonstrate, the cameras' use as vicarious agents fosters the operators' involvement and presence in the outside world, albeit in a precariously 'mediated' fashion. Whilst the barrier of the camera screen can psychologically impair the operators, being in several places simultaneously at speed, and keeping up with a particular incident, can bring a converse feeling of meaningful satisfaction:

> It was so ace catching that guy, as we'd to follow him across the city centre on the cameras as he was dodging in and out of side streets and shopping centres. He thought he'd given us the slip as well. Gave us a right buzz.
>
> (Operator 22)

Thus, not only does surveillance enable and constrain, so too, it would appear, do CCTV cameras and screens (Lyon 1994).

This section has outlined the multitude of ways the human spirit resists and escapes domination in the workplace by locating and creating within technologies of security covert opportunities and unofficial practices. Such breaks from reality, as Cohen and Taylor (1992) note, are meaningful only in so far as they are *temporary*, as prolonged overindulgence serves to weaken the escape's overall significance (function) and to risk further employment sanctions (consequences). In this sense, they are fragile and negotiated. Indeed, in order to be sustained, they must be subtle and succinct, valuable but discreet. Secondary adjustments help integrate disillusioned and alienated workers and return to the job a sense of balance, meaning and satisfaction. As such, their purpose is re-empowerment, their motivation liberation.

CONCLUSION: REASSEMBLING THE ROLES

This chapter has brought fresh theoretical and empirical insight to an issue of central importance in surveillance studies – the indefinite and ambiguous ontological nature of power within technologies of security, particularly as seen through the multiplicity and diversity of roles adopted by those who operate the systems. CCTV operators are not simply robotic watchers, they are also workers and human beings, made up of an assortment of contrasting and changing identities, subjectivities and emotive states of being. It has been argued that the actualities of CCTV monitoring are complex and that operators experience continual role conflict in their everyday lives at work. This takes the form of an ontological struggle between being a watcher, a worker and a human, the contextual situation deciding which identity precariously dominates. Indeed, it is the very roles performed by operators which *simultaneously* cast them as instruments, subjects and creators of power. They have power yet experience powerlessness, they are both controlled and controllers. Operators, however, have a capability all too often overlooked in surveillance studies and in the social sciences more generally. They possess a dynamic human spirit which allows for the innovative acquisition of temporary re-empowerment through subtle and creative interactional manipulation of the system's properties (Latour 2005). It is here that a central contradiction is revealed: the fact that surveillance mechanisms are built by 'earthlings' and have (and always will have) at their centre not simply unilateral power, rationality and automation but rather a collectivity of emotionally active, interpretative human agents, capable of rule setting, rule following *and* rule resistance.

The contradiction does not end there. Indeed, the chapter has also brought attention to a profound security/insecurity dialectic, elemental to the composition of surveillance in general and to the operation of CCTV in particular – i.e. the subtle ways in which surveillance systems are generated by, and thrive upon, the production of human *emotion*, particularly fear, anxiety and uncertainty. Whilst installed as rationalised technologies of security to order, control and protect, such mechanisms, perhaps ironically, create feelings of unease within those subject to their gaze (citizens) and, as we have seen, within those exposed to their reflection (operators). Indeed, the increasing introduction of surveillance systems into every physical and virtual social space subliminally conveys as much the precariousness of safety and order as the actuality of danger and disorder, with ever more sophisticated socio-technological devices capturing and portraying the 'realities' of chaos and suffering, thus driving, through a self-fulfilling prophecy, a psychological, social and economic demand for other such apparatus. The central point is that technologies of security produce ontologies of *insecurity* – their very existence makes real the fragmentation, inequality and uncertainty of everyday life. Paradoxically, the machine (response) is as much the problem as the solution.

References

Albrechtslund, A. and Dubbeld, L. (2005) 'The plays and arts of surveillance: studying surveillance as entertainment', *Surveillance and Society*, 3(2/3): 216–221.

Andrejevic, M. (2004) *Reality TV: The Work of Being Watched*, Oxford: Rowman and Littlefield Publishers.

Beck, U. (1992) *Risk Society: Towards a New Modernity*, London: Sage.

Biressi, A. and Nunn, H. (2005) *Reality TV: Realism and Revelation*, London: Wallflower Press.

Cohen, S. and Taylor, L. (1992) *Escape Attempts*, London: Routledge.

Dubbeld, L. (2003) 'Observing bodies. Camera surveillance and the significance of the body', *Ethics and Information Technology*, (5): 151–162.

Foucault, M. (1972) *Archaeology of Knowledge*, New York: Pantheon.

Foucault, M. (1977) *Discipline and Punish: The Birth of the Prison*, London: Penguin Books.

Foucault, M. (1980) *Power/Knowledge*, Brighton: Harvester.

Fyfe, N.R. and Bannister, J. (1996) 'City watching: closed circuit television surveillance in public spaces', *Area*, 28(1): 37–46.

Giddens, A. (1984) *The Constitution of Society: Outline of the Theory of Structuration*, Cambridge: Polity Press.

Giddens, A. (1990) *The Consequences of Modernity*, Stanford: Stanford University Press.

Goffman, E. (1961) *Asylums: essays on the social situation of mental patients and other inmates*, New York: Doubleday.

Goffman, E. (1967) *Interaction Ritual: Essays in Face-to-Face Behaviour*, Chicago: Aldine Publishing Company.

Goffman, E. (1971) *Relations in Public: Microstudies of the Public Order*, New York: Basic Books.

Hill, A. (2005) *Reality TV: Audiences and Popular Factual Television*, London: Routledge.

Hochschild, A. (1979) 'Emotion work, feeling rules and social structure', *American Journal of Sociology*, 85(3): 551–575.

Hochschild, A. (1983) *The Managed Heart: Commercialization of Human Feeling*, Berkeley: University of California Press.

Latour, B. (2005) *Reassembling the Social: An Introduction to Actor-Network-Theory*, Oxford: Oxford University Press.

Lockwood, D. (1964) 'Social integration and system integration', in Z. Zollschan and W. Hirsch (eds) *Explorations in Social Change*, London: Routledge and Kegan Paul.

Lomell, H.M. (2004) 'Targeting the unwanted: video surveillance and categorical exclusion in Oslo, Norway', *Surveillance and Society*, 2(2/3): 346–360.

Lyon, D. (1994) *The Electronic Eye: The Rise of Surveillance Society*, Minneapolis: University of Minnesota Press.

McCahill, M. (2002) *The Surveillance Web: The rise of visual surveillance in an English city*, Cullompton: Willan Publishing.

McCahill, M. and Norris, C. (2003) 'CCTV systems in London: Their structures and practices', Working Paper No. 10, Urbaneye Project, http://www.urbaneye.net/results/ue_wp10.pdf (accessed 8 October 2007).

McGrath, J. (2004) *Loving Big Brother: Surveillance Culture and Performance Space*, London: Routledge.

Neyland, D. (2004) 'Closed circuits of interaction?', *Information, Communication and Society*, 7(2): 252–271.

Neyland, D. (2006a) *Privacy, Surveillance and Public Trust*, London: Palgrave Macmillan.

Neyland, D. (2006b) 'Moving images: The Mobility and immobility of "kids standing still"', *Sociological Review*, 54(2): 363–381.

Norris, C. and Armstrong, G. (1997) *The Unforgiving Eye: CCTV Surveillance in Public Space*, Hull: University of Hull.

Norris, C. and Armstrong, G. (1998) 'Power and vision', in C. Norris, J. Moran and G. Armstrong (eds) (1998) *Surveillance, Closed Circuit Television and Social Control*, Aldershot: Ashgate.

Norris, C. and Armstrong, G. (1999) *The Maximum Surveillance Society: The Rise of CCTV*, Oxford: Berg.

Smith, G.J.D. (2004) 'Behind the screens: Examining constructions of deviance and informal practices among CCTV control room operators in the UK', *Surveillance and Society*, 2(2/3): 376–395.

Smith, G.J.D. (2007a) 'Exploring relations between watchers and watched in control(led) systems: Strategies and tactics', *Surveillance and Society*, 4(4): 280–313.

Smith, G.J.D. (2007b) 'On a different frequency? Exploring tensions between agents of control within the night-time economy', in R. Atkinson and G. Helms, *Securing an Urban Renaissance: Crime, Community, and British Urban Policy*, Bristol: The Policy Press.

Chapter 7

Hijacking surveillance?

The new moral landscapes of amateur photographing

Hille Koskela

Vignette I

In October 1997 two police officers were shot dead in the southern city centre of Helsinki. The crime investigation was conducted by the police. At that time, the police themselves did not have surveillance camera coverage in the area where the shooting took place. However, as part of the investigation the police were asking for tapes from various groups in the area, such as the army headquarters, private companies and residential houses. The huge number of tapes they received was beyond all expectations.

Vignette II

In October 2006 two security guards beat and kicked a man lying down on a street in Kontula, an eastern suburb of Helsinki. The crime was videotaped with a mobile phone camera by a man who was passing by. This man then uploaded the videotape to YouTube under a pen name 'AllSeeingEyez'. The tape caught the attention of the Finnish media. Consequently, the security company temporarily sacked the guards and formally asked the police to start a crime investigation.

These two simple vignettes are telling examples of what has been happening in the field of surveillance during the past ten years. In both cases, surveillance has a central role, yet they look very different. Vignette I taught Finns (even the police themselves) that surveillance was widespread but no one knew much about it, not to speak about having a camera register or other forms of legal regulation of surveillance. The cameras were treated as 'eyewitnesses': they happened to be there and were able to tell something afterwards. The case was strictly led by the authorities. It was easy to identify 'the good' and 'the bad'. In vignette II the authorities had no role whatsoever until after a long series of events: a passer-by, a mobile phone, YouTube, public attention, media attention, and eventually the security company worried about its reputation. Only after this were the police interested in the case. Individuals had a central role. First, the tape was shot by an independent bystander, then, the public debate started in the virtual community. Some individuals were fostering the debate, some just followed it. Altogether, the tape was viewed globally 172,362 times in the 11 months following the occasion. Furthermore, the case was fuzzy: the guards who are supposed to protect the public were committing a crime, a passer-by who is supposed to be protected as a member of the public took care of the surveillance work, and the police appeared as outsiders until the last moment.

The development of surveillance has been quite similar in most western countries: technologies are introduced first, laws are enacted afterwards, and the authorities have difficulties in 'maintaining credibility' (Hannah 1997: 175; see also Oc and Tiesdell 2000). It seems that the authorities cannot control how and where surveillance is used. At the end of the 1990s Reg Whitaker (1999: 134) claimed that 'the one-way transparency sought by the Orwellian state has been realized much more effectively in the private than in the public sector'. From then on, the use of surveillance technologies has slid from the private *sector* to private *individuals*. I call this trend 're-privatisation of surveillance'. Changes take place step by step, but the direction is clear: technology is used in ever smaller units and its distribution has become ever freer. The practices of social monitoring have become dispersed and overlapping (Ball and Webster 2003; Huey et al 2006). Wherever new forms of technology arise, new forms of resistance and alternative implementation are created.

In this chapter I will focus on the increasing amateur participation in surveillance: counter-surveillance practices, private webcams, the media use of private mobile phone photographs, and the presentation of amateur pictures and videos on internet sites such as MySpace, YouTube and Flickr. I will discuss the new moralities surrounding this phenomenon, asking how this development will 1) change how we should theorise surveillance in the future, 2) modify social relations and moralities and 3) change the political positions of 'the authorities' and 'the public'. My argument is that people's reactions to surveillance have reached a new step which is beyond passive acceptance or critical debate, even beyond purpose-oriented counter surveillance. I call this phase 'hijacking surveillance'. Many of the practices which sustain this condition can be classified as arbitrary observations rather than conscious surveillance. Still, the change is powerful and widespread and challenges the conventional ethics of seeing and being seen, of presenting and circulating images.

'EVADING DEFINITIONS' – SURVEILLANCE THEORY REVISITED

As the vignettes show, in the mid 1990s surveillance was mainly characterised as something conducted by the authorities. From that moment to the present day, there has been a fundamental change. People are increasingly participating in the production of surveillance in their everyday lives. This change has been enabled by the proliferation of information and communication technologies, especially by two lines of development: new *equipment* – namely, easily accessible home surveillance devices, webcams and mobile phones with cameras – and new *arenas* – global communities in the virtual space of the internet. An essential consequence of this condition where everybody is able to photograph or videotape nearly anything anywhere is that it has become more and more difficult to define what 'surveillance' actually is. As Kevin Haggerty (2006: 39) points out, when discussing the complexity of surveillance, '[i]t has become profoundly difficult to say anything about surveillance that is generally true across all, or even most, instances'. The limits of surveillance are evading definition.

By now, most researchers, politicians and surveillance authorities agree that surveillance has run out of sight of any organised control. In the late 1990s the critique of 'traditional' surveillance focused, for example, on the changes it might cause in space and social practices. It was presumed that surveillance mirrors fears about populations regarded as different and ensures exclusion of delinquency and deviance (e.g. Graham 1998; Fyfe and Bannister 1998; Oc and Tiesdell 2000). Surveillance was regarded as a 'powerful tool in managing and enforcing exclusion' (Norris 2002: 267). It was also believed that electronic means will increasingly replace informal social control and create a public feeling that *there is no longer need to watch over each other* (e.g. Allen 1994; Oc and Tiesdell 2000; Fyfe and Bannister 1998). Foucauldian thinking, according to which the exercise of disciplinary power 'involves regulation through visibility' (Hannah 1997: 171), was the dominant paradigm, and it was somewhat uncritically thought that being visible (i.e. under surveillance) equated with being overpowered. When the public is seen, knowledge can be collected, discipline exercised, and the 'zones of disorder' (Foucault 1980: 153) eliminated. It was claimed that visibility helps to ensure (social) purity and keep (social) space clean; that '[v]isibility is cleanliness: "light" equates with "soap" ' (Koskela 2000: 260). Being conscious of being watched by invisible overseers was understood as leading to internalisation of control. Often surveillance was considered as something inherently negative, a state in which the public is unfairly placed under a 'constant *torture* of the random but ever possible gaze' (Ainley 1998: 90, italics added). When the process of control is internalised, it seems indifferent who is watching – 'the control does not depend on who is responsible for it' (Koskela 2003: 302).

One crucial change came along with the *digital turn*: surveillance became more subtle and intense, and computer-integrated surveillance systems linked visual surveillance to other forms of technological control (e.g. Graham 1998; Green 1999; Whitaker 1999; Lyon 2001). The global arena of the internet moved 'surveillance integration' to a new level (Lyon 1998). While old surveillance watched over an anonymous crowd, the new one, at its best, could recognise individuals and combine faces to databases of criminals, activists, etc. New technologies, as Whitaker (1999: 140) points out, 'render individuals "visible" in ways that Bentham could not even conceive, but they are visible to multiple gazes coming from many different directions looking for different things'. What followed was a *rhizomatic expansion* of surveillance (Haggerty and Ericson 2000) in which it was possible to gather together crumbs of information from here and there. Digitalisation enabled 'social sorting' and intensified 'the ability to store, sort, classify, retrieve and match which is all important' (Norris and Armstrong 1999: 219).

Nevertheless, surveillance is still often conceived as something conducted by the authorities and the complex power relations remain vaguely theorised. As Laura Huey, Kevin Walby and Aaron Doyle (2006: 149) have accurately argued, surveillance:

> . . . is usually conceptualized as an activity engaged in by elites for purposes of controlling subordinate social classes. Indeed, the usual understanding of the term **surveillance** is of an omnipresent, omnipotent, and centralized political apparatus keeping tabs on citizens.

Little by little, this definition has been challenged and post-Foucauldian thinking is changing the paradigm of surveillance studies. Surveillance has ceased being centralised and spread around to directions which ten years ago were yet to be imagined. Internalisation of discipline is accompanied by creative, empowering ways of being undisciplined. The digital is not only important for the authorities and elites, but carries with it an emancipatory potential. It forms 'complex networks of power relations and resistances' (Green 1999: 27). Previously, it was argued that the mass of surveillance data would be so huge it would be 'impossible to handle' (Lyon 2001: 52) and hence 'useless'. New ways of using visual material require redefinition of usefulness or uselessness because the possible 'uses' have been multiplied. It was also argued that surveillance involves 'the quest for information' (Marx 2002: 17) and that the power of documentary accumulation depends fundamentally on the ability to make classifications. The new forms of surveillance do not necessarily aim for classifications, but in spite of this they can be powerful.

Since Thomas Mathiesen (1997) wrote about 'synopticism', it has become clear that more and more people are constituted as viewers. Surveillance is omnipresent, not only directed at the poor or marginalised. In surveillance politics 'both "sides" submit claims' (Haggerty 2006: 33), not only the ones who were traditionally perceived as being in control. The normative orientation of the analysis of surveillance is less clear than it was previously thought. New developments should not only be read negatively, but some surveillance practices 'might be accepted as a positive development' (ibid: 35). By televisualisation, cyberspace distribution and counter-observation, surveillance ends up being quite open (Koskela 2004). Performativity is an essential part of surveillance and hence the discussion about surveillance should not be reduced to the debate between crime control and privacy rights (McGrath 2004).

Rather than being confined to the ethos of discipline, power has become dispersed and flexible. Kirstie Ball and Frank Webster (2003: 12) ask for 'the paradoxical nature of surveillance' to be recognised, pointing out that 'it intrudes and enables one at the same time'. People 'participate in formerly centralized forms of surveillance and verification' (Andrejevic 2007: 222). While some of the new forms of control are increasingly involuntary (Marx 2002), it is also evident that many people are eager to participate in new technologies which involve various forms of (self)control (e.g. Knight 2000; Frohne 2002; Wise 2004). The relations of surveillance 'have been entrenched, transformed and reversed by the varied deployment of information technologies in the virtual and material worlds' (Green 1999: 42). People are not just passively adjusting to surveillance but taking *active* roles in producing and circulating images (Tinic 2006).

The production of visual material takes new forms, being more interactive and global, less concerned about privacy or tracking, but also more commercialised. Haggerty (2006: 29) presents these changes in a nutshell:

> The multiplication of the sites of surveillance ruptures the unidirectional nature of the gaze, transforming surveillance from a dynamic of a microscope to one where knowledge and images of unexpected intensity and assorted distortions cascade from viewer to viewer and across institutions, emerging in unpredictable configurations and combinations, while undermining the neat distinction between watchers and watched through a proliferation of criss-crossing, overlapping and intersecting scrutiny.

'SPLINTERING PICTURES' – RETHINKING IMAGES AND FACTS

As a consequence of this dispersed scrutiny, the moral binary opposition good/bad is splintering and previously clear categories are breaking. The differences between the authorities and the public, outsiders and insiders, the controlled and the controllers, have become less clear. New forms of scrutiny are spreading, making the old story about the 'good police officers' chasing 'evil criminals' sound like a naive fairy tale. Yet, simultaneously, perceptions about counter-surveillance by the public are often idealised. Both traditional surveillance and voluntary vigilance may have consequences which are either good or bad, or anything between. Shedding light on the splinters warrants conceptualising 'a reimagined relationship between police authorities and the community they serve' (Huey et al 2006: 150). Since 'myriad agencies now trace and track mundane activities for a plethora of purposes' (Lyon 2002: 13), risks of misuse and misinterpretation increase. While traditional surveillance systems were 'leaking' and surveillance tapes ended up in wrong hands – as the cases of sexual harassment with surveillance cameras prove (e.g. Hillier 1996; Koskela 2002) – new forms of scrutiny may multiply these kinds of problems. Another challenge is formed around the question of who is/are able to define what is misuse and what is appropriate use.

But there is more to this than just new practices. Transition from modern world to post-modern times has meant that solid knowledge is replaced by splintering world views and differing interpretations. Relativistic views challenge the notion of 'the objective', proving that most things, arguments and images are actually socially produced and politically loaded. People are actively involved in the interpretation of images and signs, and socially produced connotations have replaced objective facts. As Peter Weibel (2002: 219) has crystallised: 'We live in a society that prefers the sign to the thing, the image to the fact.'

'[T]he illusory power of representation' (Balshaw and Kennedy 2000: 3) is intertwined with the tricky question of 'truth'. One of the key explanations for the popularity of live visual representations lies in 'the fetishization of real time and live-effects' (Frohne 2002: 256). Pascal Pinck (2000) writes about *the skycam* – a news helicopter reporting crime to television. He claims that instability of the image, action and movement, as well as spatial proximity, are stylistic measures which are used consciously in order to create an impression of reality. There is a historical continuum of what has been perceived as a 'proof for truth'. In the late nineteenth century, photography was somewhat uncritically perceived as proof. As J. Macgregor Wise (2004: 424) points out: '[E]arly photography contributed to a modern regime of truth: the camera would reveal the world in a new way.' Since the late twentieth century it has been clear that photographs can be manipulated, characters erased, and that pictures are not evidently true. TV took the place of photography and, in Pinck's (2000: 64) words, 'we are only likely to believe what we see on TV'. Presently, the truth value of television has faltered. Reality TV shows have not contributed to giving the TV impression of being more real – on the contrary. While the 'fiction' has increasingly become 'indistinguishable from reality' (Zizek 2002: 226), ever more proof is needed that anything appears true.

Surveillance videos have taken over the place of television. In the slippery field of representations, surveillance material has remained quite solid. The old 'flickering piece of black-and-white footage' (Doyle 2006: 199) is the icon of surveillance imaginary. As John McGrath (2004: 52) argues: 'The viewer of surveillance footage is expected to relate to the footage as self-evident description of the events recorded.' Surveillance is perceived as something mainly conducted by the authorities and hence is often inherently trustful, but counter-surveillance – or any material which, for example, on the internet endeavours to be construed as true reality – has purposefully used same connotations. Even so, it is clear that the 'reality' of a videotape is a social product rather than a mere description. As Maria Balshaw and Liam Kennedy (2000: 8) emphasise, 'the operations of the eye are not only biological and formal, but also cultural and psychological'. Not a single everyday 'text' – such as a surveillance video on TV, or a mobile phone snapshot in a tabloid – is explained in a universally coherent way, but different audiences provide different readings of it. Surveillance – or counter-surveillance – 'does not find knowledge, but creates it' (Allen 1994: 144). All visual images are inherently political. In spite of this, the role of the citizens – what they are asked for as well as one of their spontaneous motivations – still is to *verify* and to gather *evidence*.

THE EVERYDAY SYNOPTIC

The political nature of images means that there is a need to re-think not only surveillance material produced by the authorities but also the myriad new ways of amateur photography which contribute in producing images which 'enter' the field of surveillance. In the academy, there has been quite a lot of reasoning about the *everydayness* of surveillance (Haggerty and Ericson 2000; Staples 2000; Lyon 2001; Ball and Webster 2003). This everydayness, however, does not merely mean that people are increasingly controlled by sophisticated devices or that they leave traces and are increasingly trackable. The other side of this development is that people increasingly *have access to* what can be described as surveillance technologies. The 'embeddedness' of surveillance in the everyday thus realises in people's lives in two ways: both as intensified control (by others) and as active agency in producing control (by themselves). This dual change does not necessarily lead to more control, but to more images which can be used for manifold purposes. As I shall argue below, these purposes range from random arbitrary witnessing to purposeful politics of resistance. People resist 'organisational forms of power by surveillance activities' (Huey et al 2006: 149). Surveillance has fundamentally changed from the 'centralized political apparatus' to a practice to which anyone can contribute. The old-fashioned surveillance systems are still in use but they are accompanied by various voluntary surveillance practices.

Mark Andrejevic (2007: 212) describes this development as 'digital enclosure':

> Within the digital enclosure, the movements and activities of individuals equipped with interactive devices become increasingly transparent – and this makes monitoring technologies easier to obtain and use. The result is increasing public access to the means of surveillance – not just by corporations and the state, but by individuals.

Citizen participation in surveillance is by no means a new phenomenon. While the genre of 'reality TV' is widespread altogether, from the very beginning one of its roles has been to enable citizens to participate in crime prevention and control – to be engaged in surveillance (Korander 2000; Pinck 2000; McCahill 2003, among others). According to McGrath (2004: 29), interactive crime prevention programmes endeavour 'to involve the public in a visceral engagement with the complexities and sometimes brutalities of modern policing'. Surveillance and the mass media have an unholy alliance. As Clive Norris and Gary Armstrong (1999:67) argue, television and CCTV 'were made for each other'. Crime-prevention programmes have used surveillance-camera tapes and surveillance-based simulations to create an impression of the public being able to verify crime and participate in police work. The audience is able to identify unknown suspects shown in the 'video wanted posters' (Doyle 2006), and interactivity is guaranteed by 'hot line' telephone numbers and internet sites. Voluntary image production can be conceived as being a continuation of this trend.

Let me reverse a bit and re-think the point I made previously about the two lines of development: new equipment and new arenas. The everydayness means, among other things, that one can buy a surveillance camera from a hardware store or a webcam from a stationery shop. Then again, since cameras and video recorders are installed in mobile phones, many people constantly carry a camera in their pocket. Timo Kopomaa (2000) has tellingly used the phrase 'city in your pocket' to describe the use of mobile phones. While camera phones are not literally surveillance devices, anyone with a camera phone can all of a sudden turn into an observer. With camera phones, webcams and other everyday devices, people are able to watch each other more than ever before. Furthermore, 'interactive' webcams intensify observing: the audience can watch detailed real-time videos, pan and zoom, and try to recognise people walking on the street. People are also easily able to use digital video recorders or 'traditional surveillance cameras' for their own purposes – to watch their houses or other people they come in contact with. Altogether, people are creating emerging ways of seeing, viewing others and representing. The photographic act 'is becoming more common and more commonplace' (Rivière 2005: 181). The equipment is there.

There are endless possibilities 'to release videotaped footage to the news media' or to 'download streams of images onto the internet' (Huey et al 2006: 154). The new devices allow individuals to create 'shows' without 'outside editors, directors, or producers to decide who gets how much airtime' (Knight 2000: 24), giving them an impression of '[t]he constant *potential* for breaking news' (Pinck 2000: 65). The commercial media is reinforcing this by encouraging people to observe any odd or newsworthy occasions by buying and publishing pictures (often specifically shot by a mobile phone) or videos. The mass of presentations of amateur images on internet sites such as photoblogs, Flickr, MySpace or YouTube is huge and these sites are easy to access. New technologies afford users the ability to re-present and perform the everyday and provide them with the possibility of self-expression in global media. The arena is there.

MORAL LANDSCAPES

The Möbius Strip

Nevertheless, to have access to an arena is not enough. People need reasons for using the new technologies. They need *motivations*. The question of motivations intertwines with the idea of *moral landscapes* like the Möbius Strip – a never-ending continuous curve one side of which cannot be distinguished from the other. What matters is not that people are able to use the equipment and enter the arena, but the questions surrounding their motivations: how and why do people use these facilities? Eventually, what is shot and circulated is largely independent of the device: all technologies can be used with many different ends in view. Topic is of utmost importance and deserves a closer look. As in the case of surveillance in general, it is evident that amateur photographing or videotaping do not provide 'innocent' illustrations of material space but have both deliberate and unintended consequences. Visual technologies can embody a new regime of order but they can also be used as new means of empowerment (Koskela 2004). They can serve surveillance but, reciprocally, also liberation, resistance or escape (Bogard 2006).

Watching and being watched is structured around reveries of voyeurism and exhibitionism (Tabor 2001). There is a voyeuristic fascination in looking but, reciprocally, an exhibitionist fascination in being seen. Visual representation 'mediates scopophilic and voyeuristic desires (to look, to be seen)' and 'technologises the act of seeing (the fusion of the eye and the camera lens)' (Balshaw and Kennedy 2000: 7). In post-modern societies, this seems quite natural: 'Surveillance can become spectacle and the people can enjoy surveillance as a spectacle because seeing is entangled with sexuality and power' (Weibel 2002: 219). People are, indeed, involved in 'getting pleasure from viewing the forbidden, or in viewing without being viewed' (Doyle 2006: 212). The voyeuristic nature of traditional surveillance is justified by its protective nature. How about voluntary vigilance?

Counter-surveillance activism

Political reactions, artistic presentations and activist interventions have followed surveillance from the beginning. Often these fuse, making it quite hard to divide art from activism or vice versa. The debate has mainly resided in two opposing arguments: some critics have gone for resisting surveillance and endeavouring to make it more transparent and less intensive, others believe that the best way to channel critique and support democracy is to intensify surveillance coming 'from below'. Resisting surveillance includes hiding from it, providing the public opportunities to hide, or demanding tighter regulation. Intensifying it includes taking possession of surveillance equipment and using it for 'alternative' purposes.

The definition of 'counter-surveillance' is far from solid. Torin Monahan (2006: 515) describes it as 'intentional, tactical uses, or disruptions of surveillance technologies to challenge institutional power asymmetries' (see, however, McGrath 2004 for a much wider definition). Critical debate and resistance of surveillance is organised by non-governmental organisations such as American Civil Liberties Union, Electronic Frontier Foundation, Omega Foundation and Privacy International, which effectively reveal misuses and unbalanced power structures. For example, Privacy International and a number of affiliate human rights groups throughout the world present 'Big Brother awards' to the government or private-sector organisations which have excelled in the violation of privacy in their countries.

Even more creative critique is provided by art groups such as Surveillance Camera Players which stages public plays to draw attention to the prevalence of surveillance in society (Monahan 2006). Institute for Applied Autonomy (IAA) has been one of the pioneers for organising 'surveillance-free zones'. The project iSee provides surveillance-free routes around cities, allowing people 'to play a more active role in choosing when and how they are recorded' (Schienke and IAA 2002). The project endeavours to raise public awareness of surveillance and to criticise any control that undermines civil liberties. IAA has also developed creative ways of using surveillance technologies 'for activism' against the police who try to control public demonstrations (IAA 2006). The group claims that increasingly 'citizens have appropriated information and communication technologies to invert the power relations embodied by traditional surveillance regimes' (ibid.: 168). They encourage activists to use cameras in street protests in order to produce photo and video documentation of police activities, aiming to both 'mitigate' police behaviour and to provide documentary evidence. Citizens with 'anti-police values' (Huey et al 2006: 155) have employed a model of 'defensive surveillance' which embodies 'a tactic of speed' (IAA 2006: 173). It allows the actor of surveillance to leave the field whenever necessary, hence undermining the logic of force.

A well-known example of slightly different counter-surveillance is Steve

Mann's aim to 'examine how using wearable computing devices can promote personal empowerment in human/technology/human interactions' (Mann et al 2003: 336). Wearable surveillance equipment enables its users to watch the watchers and to criticise their practices. It distributes images via the internet as the user walks around wearing it, hence integrating the global audience to the project. Whenever questions are asked about the event, neither the user nor the security guards targeted 'know how many copies of my transmitted pictures might have been made' (Mann 2002: 535). Mann calls his philosophical framework 'reflectionism', which means 'turning those same tools against the oppressors' (Mann 2002: 534). Terms such as 'sousveillance' and 'co-veillance' are also used to describe the agency which the public gains with the equipment. Mann's projects challenge the rhetoric of public safety embedded in surveillance, the unquestionable nature of the authorities and the criminalisation of the critic of surveillance. Further, the projects highlight the embodied experience of observing and being observed and, in doing so, disrupt 'the illusion of detached, objective, impersonal, disembodied monitoring' (Monahan 2006: 524).

Another, more straightforward example of counter-surveillance is *cop watching*. Since the widespread public attention surrounding the beating of Rodney King, 'the police cannot be at all sure that any act of brutality has not been recorded and passed on to broadcasters' (McGrath 2004: 199). Volunteers organise movements which aim to reveal police brutality and to place the 'authorities under scrutiny through the use of camera surveillance equipment' (Huey et al. 2006: 162). Cop watching is a way of monitoring which serves a broader context of political action. As Huey and her colleagues (ibid.: 150) point out, the members of cop-watch organisations see their work as ultimately democratic and just – 'as promoting democratic accountability of a state institution that has tremendous power in the lives of marginalized citizens' – and tend to frame their work in the language of democracy.

From emphatising to bullying

All voluntary vigilance cannot be defined as counter-surveillance. Interactive photography can simply be a way of *emphatising* – of 'assimilating a sensitive relationship to the world and to other people' (Rivière 2005: 181). New technologies are used in order to capture and archive daily life. For some, the 'notion of self-ownership' (Mann 2002: 533) is enough of a reason. Images are encountered through a number of registers – 'the discursive, the bodily, the sensory, the psychic and the emotional' (Rose 2004: 551). These encounters stand at the side of the purpose-oriented political motivations. Some forms of photographing refuse to serve 'resistance'. Simply, 'the camera's gaze . . . reveals the world in a new way, reveals aspects of us that we are unaware of (habits, expressions) and contributes to new social formations' (Wise 2004: 425).

Images are used as a form of interpersonal communication. Taking and sharing photographs 'creates the perception of "being together" founded on an active reality' (Rivière 2005: 183). Sometimes, representations are socially more important than the original experiences. They promote 'ephemeral, fluctuating modes of belonging' (ibid.: 183) and support collective perception. Images are 'powerful means for *doing togetherness*' (Rose 2004: 560, italics added). Webcams, for example, can be used for sending 'electronic postcards' which promote touristic collectivism and sustain memorising. New York's Times Square is perhaps the best example of this, having several interactive real-time webcams (for a more detailed discussion see Koskela 2006) with which people are able to e-mail postcards – pictures of themselves on the square. The website Times Square Cam keeps up a Hall of Fame archive consisting of copies of the thousands of pictures sent. While the pictures, ostensibly, are trite and repetitive, they are somehow fascinating in their triviality. The ordinariness of the images – despite them being highly technologised – is so intense that they become extraordinary. Sharing visual memories is not indifferent even if the practice would have no aims outside this communication.

Nevertheless, new media also have 'potential as a socially fragmenting force' (Tinic 2006: 310). As they are able to include, they provide means for exclusion. There are plenty of examples of 'internet bullying' where original or manipulated pictures or videotapes have been used as a means for humiliation of certain individuals or groups. The cases range from global scandals to local misconceptions. The two examples below highlight the two ends of this spectrum.

In March 2007 YouTube published a manipulated video presenting the US presidential candidate Hillary Clinton as a dictator-like figure in a 1984 Orwellian setting applied from an old Apple computer add. This led to a brief 'online mystery' about who was behind the video, as well as an extremely wide political debate. By the end of 2007 the video had been viewed more than 4 million times and a Google search 'Hillary 1984' gave almost 30,000 pages around the world. In May 2007 a 15-year-old schoolboy from Finland loaded a video on YouTube featuring his teacher singing at a school party. The video was entitled 'Karaoke at the mental hospital', hinting that the performer was a mental patient. Later, the boy was found in court 'guilty of libel'. There are also plenty of examples of creating defamatory mock MySpace or Facebook profiles, etc. The prejudices of physical life are reflected – and sometimes reinforced – in the virtual life (e.g. Higgins et al 1999). Amateur images can be used for nihilist purposes in ways which connect to the anxious encounters and conflicts in social life.

New moral (panic)s

An important side of the new visual technologies, which shows a parallel with the 'old-fashioned surveillance', is their connection to security. They can be – and increasingly are – used to avoid perceived risks and increase security, ranging from 'watching over' to 'moral panics'. People employ 'monitoring strategies as a means of taking responsibility of one's own security' (Andrejevic 2007: 218). As Weibel (2002: 207) argues, 'absolute visibility is legitimated with the claim and guarantee of absolute security'. At the moment, the insecurities people face are more difficult to grasp than ever before. Zygmund Bauman (2006: 130) talks about 'security obsession'. Caution, mistrust and tense social relations form a new condition, the 'culture of fear' (Furedi 2002), which is characterised by 'the continuous reformulation of ordinary experience as dangerous' (ibid.: 113). Fear as a situated experience and fear as a transformation of visual culture are increasingly connected.

Largely, the new importance of security has come along with a neo-liberal rule which efficiently defines 'dangerous' spaces, activities, groups or individuals. Security has become 'the justification for measures that threaten the core of urban social and political life' (Marcuse 2004: 275). Social integration has been replaced by zero tolerance and inclusion by exclusion. Consequently, fear seems to provide people with a justification for protecting themselves and being vigilant. To capture images (of almost anything/anybody) can be explained within the rhetoric of security. The media is the main 'symbolic battleground' for fear and security, whether they would be local or global. As Ursula Frohne (2002: 255) incisively claims, 'the media becomes the new disciplining authority'.

The security-oriented vigilant audience makes use of the new technologies. For example, webcams in urban space give parents tempting opportunities to observe their teenage children hanging around in the city. The city of Jyväskylä, in central Finland, gives an example of what is available: its internet site contains exceptionally detailed real-time videos from the city-centre pedestrian area where teenagers are in the habit of meeting. People walking in the area can easily be recognised. A more striking and internationally meaningful example, however, is the 'border watch'. In November–December 2006 there was a temporary, openly accessible website presenting the US–Mexico border, called Texas Border Watch Test Site. During the experiment, live video footage from border surveillance cameras in Texas was available for anyone who wanted to contribute to border control and exclusion of unwanted immigrants. The site proved to be popular among US residents and there are serious plans to make it permanent.

Any 'wired' device can be effectively linked to flows of information. The images can be immediately distributed worldwide and consumed by others. There is more to this than just the chance for spontaneous snapshots or videotaping. New visualities easily contribute to 'the social construction of suspicion' (Norris and Armstrong 1999: 117), creating new forms of social control. As McGrath (2004: 22) remarks: 'Every camera functions within a field of power and prejudice structured by visual markers.' Gathering knowledge is a form of maintaining control. A look – again, in a new forum, taking new forms – equates with a 'judgmental gaze' (Burgin 2002: 235). Power works through new moralities.

The amateur paparazzis

Some of the new ways of observing can be extremely macabre. A brief look at the seemingly trivial images of open webcams on the internet or mobile phone cams in tabloids is enough to reveal that a huge amount of material is published with malignant delight. People shoot potential scandals, public figures in awkward situations, or accidents in their neighbourhoods, rejoicing at other people's misfortunes. As an example, a sketchy analysis of the 100 latest readers' mobile phone pictures on the website of a Finnish free tabloid, Uutislehti 100, showed that one-third of the pictures published showed traffic accidents, fires or witnessed crime scenes. Just as crime-control TV programmes have been described as 'crime control pornography' (Korander 2000: 185) – where the viewers can, under the pretext of crime control, both moralise and peep into the field of criminal action – this amateur imagery is reminiscent of 'pornography'. This type of observation is deeply profane, as of the old-times circus audiences watching 'negros' or 'thrown dwarfs': observing 'the odd'.

Even so, this macabre genre is only one of the forms in which public scrutiny can be experienced as a pleasurable activity. A lot of amusing and ironic material is published, showing better taste. Amateur images promote the fun features and entertainment value of surveillance (Haggerty 2006) and show that 'surveillance-enabling technologies are able to perform entertainment functions' (Albrechtslund and Dubbeld 2005: 217). The contradictory desire for surveillance is also present in popular culture (Pecora 2002). There is a good amount of fun, play and hedonism included, illustrating out that '[o]bservation is not a menace; observation is entertaining' (Weibel 2002: 218).

However, since visibility is no longer necessarily interpreted as a threat, there is, among other trends, increased willingness to be seen – a deliberate exposure of the self and of private lives (Knight 2000; Jimroglou 2001; Burgin 2002). New technologies are used not only for observation but to 'prove the presence' of oneself. Private non-commercial webcams present the daily lives of individuals, often promoting nothing but the existence of the person presenting themselves, taking the viewers 'from here to banality' (Mosco 2005: 18). When 'the camera moves from a recording instrument to an integral aspect of the subjects' lives' (Knight 2000: 24), exposing oneself is connected to identity formation. As Slavoj Zizek (2002: 225) argues: 'Today, anxiety seems to arise from the prospect of NOT being exposed to the Other's gaze all the time, so that the subject needs the camera's gaze as a kind of ontological guarantee of his/her being.' Contemporary fear is 'that of being withdrawn completely from the gaze of the others' (Frohne 2002: 275) and there is a 'compulsive desire to attain tele-presence, to verify and validate one's own existence' (ibid.: 256). People wish to have agency in the era of surveillance. Indeed, they can be subjects rather than objects of watching, which can be liberating and confidence-bolstering (Knight 2000), and empowering (Koskela 2004). As I have previously argued (ibid.; 2006), revealing can be conceptualised as a political act – a way of resisting the objectifying nature of surveillance.

It must be noted, however, that often individual watching or presenting practices are tied into the process of *consumption*. The digital economy offers not only a forum for independent image circulation but also new marketing concepts. Frohne (2002) talks about the 'economy of attention'. Indeed, one important binary opposition is structured around the pair activism versus consumerism – 'the commercial deployment of interactivity as information-gathering strategy' (Andrejevic 2007: 213). This includes both the commercialisation of counter-surveillance (by the media) and the harnessing of private observers to serve the purposes of the powerful (by the authorities, police, etc.). People are, more and more, in various ways, *encouraged* to observe and inform, to take part in crime prevention and control, to be committed. This development is parallel to what has been happening in the field of traditional surveillance: increasingly, monitors are turned towards the audience, hence almost forcing people to play roles in maintaining control. There is a '*rising culture of informing*' (Doyle 2006: 202, italics added), in both official and unofficial fields of observing. New forms of observing, together with their commercialisation, create a temptation for 'paparazzi mentality'. In and through the images, crimes, scandals and accidents become *commodities* (cf. Presdee 2000). The paparazzi mentality is then likely to lead to an atmosphere of mutual distrust and suspicion.

Invisible zones

One effect of new technologies becoming available and accessible is that the conventional codes of what can and what cannot be shown have been changing. Photographing devices are present in places which previously remained unphotographed: on trivial occasions and at unexpected incidents. This reveals cultural tensions surrounding epistemological conceptions of vision, and raises questions regarding the role of technology in the representation and construction of subjects (Jimroglou 2001). To discuss the *limits* of image production and circulation seems to be out of fashion at the moment. Nonetheless, some of the old-fashioned ethical questions should be revisited in the contexts of new technologies and moralities. The unintended consequences of using the new equipment 'must be weighed against potential benefits' (Huey et al 2006: 165) – and, whenever the question of benefit arises, there is a need to ask 'whose benefit?'.

New forms of observing do not necessarily include 'quest for information' – rather the material they produce could be described as *visual waste* or *information trash* (definitions which, indeed, would apply to most traditional surveillance material). The economy of attention plays a surprising role here: while information may not have been requested, it may still have been produced unintentionally. There is news which 'only became notorious because dramatic surveillance footage was broadcast' (Doyle 2006: 205). The murder of two-year-old James Bulger in Liverpool in 1993 is perhaps the most horrific example of a 'CCTV-famous' crime. The most shocking element of this case was that 'the camera was powerless to intervene in what it was witnessing' (McGrath 2004: 36). Likewise, the amateur paparazzis produce pictures which are 'potential news'.

Along with the development of technologies, the definition of *newsworthy* has changed. Previously, newsworthy issues were categorised as focusing on the unusual rather than the normal, the dramatic rather than the mundane, and the simplistic rather than the complex (McCahill 2003). New technologies have brought the mundane, normal and complex to unexpected fame. This type of dramatised representation in the media has a 'hypnotic fascination' (Miles 2003: 52). As Weibel (2002: 209) states:

> The principal structure of the regulation of visibility and invisibility refers to rejection, not only as it is registered within paranoia – although there especially – but in the entire social order. The visible field is a field of symbolic order, and just as rejections are necessarily arrived at in the symbolic order, the field of visible necessarily arrives at invisible zones.

CONCLUSIONS – HIJACKING SURVEILLANCE

In this chapter I have explored the emerging practice of amateur surveillance. It is possible to recognise the 'scale' along which the surveillance scene has slid. First, there was *passive acceptance* of surveillance, characterised by the naive, optimistic, often expressed public attitude 'I have nothing to hide'. Second, the *critical approach* arose, fostering public discussion, research projects in social studies, and surveillance-critical attention in the media, art circles and NGOs. Third, various *counter-surveillance* practices were developed by vigilant individuals, NGOs and artists. Presently, the scale has reached the fourth phase, which I call *hijacking surveillance*. People use various items of surveillance equipment for producing visual material for their own purposes with different motivations. This does not necessarily form any critical or other statements. Surveillance is not used for political aims, it presents no claims, has no objectives and there is no organisational structure behind it. There is no agenda. Yet social consequences are widespread.

The present deeply synoptic condition has hardly any qualities which would resemble the panopticon, but it is true that watching remains 'sporadic', yet 'the threat of being watched never ceases' (Hannah 1997: 347). The random but ever possible gaze accompanies us. Everybody has turned into a potential observer, but the distinction between critical overseers and amateur paparazzis remains vague. To have more observers does not lead to having ethically just observations. The political motivations of some applications of counter-surveillance – such as revealing police brutality – are fairly easy to accept, but the practice itself does not tell us anything about the motivations of those who are conducting it. Indeed, 'it matters enormously who is actually conducting surveillance' (Haggerty 2006: 33). The hype about interactivity has been accompanied by critical notions. As Andrejevic (2007: 213) points out: 'Interactivity is becoming synonymous with asymmetrical forms of monitoring, information gathering, and surveillance.'

Let me now return to the second vignette, presented at the beginning of the chapter. It consists of two parts: part one – something happened (i.e. security guards beat a man in an unknown suburb); part two – someone captured an image (i.e. a passerby took a mobile phone out of his pocket and shot what he was witnessing). Part one would most likely not have attracted public attention. Part two made the thing happen. The image was what mattered – what created the 'reality' surrounding this casual occasion. The vignette, among other things, is able to prove how the epistemology of visual images has changed, how 'everything exists only because it is an image' (Weibel 2002: 212). If the street-level example is not enough to make the point, there is another example available. Global terrorism proves that 'reality' is closely intertwined with 'the war of signs' and that images can be mobilised for different political interests. The 9/11 World Trade Center attack literally showed that to be seen – not to hide – is what makes acts meaningful.

Presently, 'wearable devices' have developed from a phantasmagorical cyborg utopia into a mundane everyday outfit. A camera phone weighs no more than a wallet in one's pocket and hence can be 'everywhere'. On any occasion it is possible that 'I observe you observing me/others'. This everydayness is not without consequences in social relations, morals and actions. The differentiation between the watchers and the watched, which previously formed the basis of surveillance theory, has disappeared. This ostensibly trivial change is, if we look closely, quite fundamental. The historical structure of the political positions of 'the authorities' and 'the public' is fading. When there is no difference between the controllers and the controlled, all politics and ethics need to be rethought. The democratic idea of representational authority is breaking down.

Voluntary vigilance can be extremely powerful – sometimes more powerful than traditional surveillance. Yet, it needs to be conceptually structured. The mess of multiple everyday technologies and multiple individual motivations is challenging surveillance theory. It has been pointed out that 'the all-seeing' power which was essential to old surveillance theory has roots in Christian religion: '[t]he overpowering and ubiquitous eye of God can be considered as prototype of this hegemonic vision' (Schmidt-Burkhardt 2002: 18). The nature of the potential overseer is 'God-like', someone who is there and simultaneously is not: '[h]is presence, which is also an absence, is in his gaze alone' (Whitaker 1999: 34). How telling is the pen name of the man in the vignette – 'AllSeeingEyez'! We all have all-seeing eyes now. The questions remains, how do we use them?

References

Ainley, R. (1998) 'Watching the detectors: Control and the Panopticon', in R. Ainley (ed.) *New Frontiers of Space, Bodies and Gender*, London: Routledge, 88–100.

Albrechtslund, A. and Dubbeld, L. (2005) 'The plays and arts of surveillance: Studying surveillance as entertainment', *Surveillance and Society*, 3(2/3): 216–221.

Allen, M. (1994) ' "See you in the city!" Perth's Citiplace and the space of surveillance', in K. Gibson and S. Watson (eds) *Metropolis Now: Planning and the urban in contemporary Australia*, Australia: Pluto Press, 137–147.

Andrejevic, M. (2007) *iSpy: Surveillance and power in the interactive era*, Lawrence: University Press of Kansas.

Ball, K. and Webster, F. (2003) 'The intensification of surveillance', in K. Ball and F. Webster (eds) *The Intensification of Surveillance: Crime, terrorism and warfare in the information age*, London: Pluto Press, 1–15.

Balshaw, M. and Kennedy, L. (2000) 'Introduction: Urban space and representation', in M. Balshaw and L. Kennedy (eds) *Urban Space and Representation*, London: Pluto Press, 1–21.

Bauman, Z. (2006) *Liquid Fear*, Cambridge: Polity Press.

Bogard, W. (2006) 'Surveillance assemblages and lines of flight', in D. Lyon (ed.) *Theorizing Surveillance: The panopticon and beyond*, Cullompton: Willan Publishing, 97–122.

Burgin, V. (2002) 'Jenni's room: Exhibitionism and solitude', in T.Y. Levin, U. Frohne and P. Weibel (eds) *Rhetorics of Surveillance from Bentham to Big Brother*, Karlsruhe: ZKM Centre for Art and Media, 228–235.

Doyle, A. (2006) 'An alternative current in surveillance and control: Broadcasting surveillance footage of crimes', in K.D. Haggerty and R.V. Ericson (eds) *The New Politics of Surveillance and Visibility*, Toronto: University of Toronto Press, 199–224.

Foucault, M. (1980) 'The eye of power', in C. Gordon (ed) *Power/Knowledge: Selected interviews and other writings 1972–1977 by Michel Foucault*, Sussex: Harvester Press, 146–165.

Frohne, U. (2002) ' "Screen tests": Media, narcissism, theatricality, and the internalised observer', in T.Y. Levin, U. Frohne and P. Weibel (eds) *Rhetorics of Surveil-*

lance from Bentham to Big Brother, Karlsruhe: ZKM Centre for Art and Media, 252–277.

Furedi, F. (2002) *Culture of Fear: Risk-taking and the morality of low expectation*, London: Continuum.

Fyfe, N.R. and Bannister, J. (1998) ' "The eyes upon the street": Closed-circuit television surveillance and the city', in N.R. Fyfe (ed) *Images of the Street: Representation, experience and control in public space*, London: Routledge, 254–267.

Graham, S. (1998) 'Spaces of surveillant simulation: New technologies, digital representations, and material geographies', *Environment and Planning D: Society and Space*, 16: 483–504.

Green, S. (1999) 'A plague on the Panoptician: Surveillance and power in the global information economy', *Information, Communication and Society*, 2: 26–44.

Haggerty, K.D. (2006) 'Tear down the walls: on demolishing the panopticon', in D. Lyon (ed) *Theorizing Surveillance: The panopticon and beyond*, Cullompton: Willan Publishing, 23–45.

Haggerty, K.D. and Ericson, R.V. (2000) 'The surveillant assemblage', *British Journal of Sociology*, 51(4): 605–622.

Hannah, M. (1997) 'Space and the structuring of disciplinary power: An interpretive review', *Geografiska Annaler*, 79B: 171–180.

Higgins, R., Rushhaija, E. and Medhurst, A. (1999) 'Technowhores', in Cutting Edge (eds) *Desire by Design: Body, territories and new technologies*, London: I.B. Tauris, 111–122.

Hillier, J. (1996) 'The gaze in the city: Video surveillance in Perth', *Australian Geographical Studies*, 34: 95–105.

Huey, L., Walby, K. and Doyle, A. (2006) 'Cop watching in the downtown Eastside: Exploring the use of (counter)surveillance as a tool of resistance', in T. Monahan (ed.) *Surveillance and Security: Technological politics and power in everyday life*, New York: Routledge, 149–165.

Institute for Applied Authonomy (2006) 'Defensive surveillance: Lessons from the republican national convention', in T. Monahan (ed) *Surveillance and Security: Technological politics and power in everyday life*, New York: Routledge, 167–174.

Jimroglou, K.M. (2001) 'A camera with a view: JenniCAM, visual representations and cyborg subjectivity', in E. Green and A. Adam (eds) *Virtual Gender: Technology, consumption and identity*, London: Routledge, 286–301.

Knight, B.A. (2000) 'Watch me! Webcams and the public exposure of private lives', *Art Journal*, 59(4): 21–25.

Kopomaa, T. (2000) *The City in Your Pocket: Birth of the mobile information society*, Helsinki: Gaudeamus.

Korander, T. (2000) 'Turvallisuus rikollisuuden ja sen pelon vastakohtana', in P. Niemelä and A.R. Lahikainen (eds) *Inhimillinen turvallisuus*, Tampere: Vastapaino, 177–216.

Koskela, H. (2006) 'The other side of surveillance. Webcams, power and agency', in D. Lyon (ed) *Theorizing Surveillance: The panopticon and beyond*, Cullompton: Willan Publishing, 163–181.

Koskela, H. (2004) 'Webcams, TV shows and mobile phones: Empowering exhibitionism', *Surveillance and Society*, 2(2/3): 199–215.

Koskela, H. (2003) ' "Cam Era" : The contemporary urban panopticon', *Surveillance and Society*, 1: 292–313.

Koskela, H. (2002) 'Video surveillance, gender and the safety of public urban space: "Peeping Tom" goes high tech?', *Urban Geography*, 23: 257–278.

Koskela, H. (2000) ' "The gaze without eyes": Video surveillance and the changing nature of urban space', *Progress in Human Geography*, 24: 243–265.

Lyon, D. (2002) 'Surveillance as social sorting: Computer codes and mobile bodies', in D. Lyon (ed) *Surveillance as Social Sorting: Privacy, risk and digital discrimination*, London: Routledge, 13–30.

Lyon, D. (2001) *Surveillance Society: Monitoring everyday life*, Buckingham: Open University Press.

Lyon, D. (1998) 'The world wide web of surveillance: The internet and off-world power-flows', *Information, Communication and Society*, 1: 1–9.

Mann, S. (2002) ' "Reflectionism" and "diffusionism": New tactics for deconstructing the video surveillance superhighway', In T.Y. Levin, U. Frohne and P. Weibel (eds) *Rhetorics of Surveillance from Bentham to Big Brother*, Karlsruhe: ZKM Centre for Art and Media, 531–543.

Mann, S., Nolan, J. and Wellman, B. (2003) 'Sousveillance: Inventing and using wearable computing devices for data collection in surveillance environments', *Surveillance and Society*, 1: 331–355.

Marcuse, P. (2004) 'The "war on terrorism" and life in cities after September 11, 2001', in S. Graham (ed). *Cities, War, and Terrorism: Towards an urban geopolitics*, Malden: Blackwell, 263–275.

Marx, G.T. (2002) 'What's new about the "new surveillance"? Classifying for change and continuity', *Surveillance and Society*, 1: 9–29.

Mathiesen, T. (1997) 'The viewer society: Foucault's "Panopticon" revisited', *Theoretical Criminology*, 1: 215–234.

McCahill, M. (2003) 'Media representations of visual surveillance', in P. Mason (ed) *Criminal Visions: Media Representations of Crime and Justice*, Devon: Willan Publishing, 192–213.

McGrath, J. (2004) *Loving Big Brother: Performance, privacy and surveillance space*, London: Routledge.

Miles, M. (2003) 'Strange days', in M. Miles and T. Hall (eds) *Urban Futures: Critical commentaries on shaping the city*, London: Routledge, 44–59.

Monahan, T. (2006) 'Counter-surveillance as political intervention?', *Social Semiotics*, 16(4): 515–534.

Mosco, V. (2005) *The Digital Sublime: Myth, power, and cyberspace*, Cambridge, Massachusetts: The MIT Press.

Norris, C. (2002) 'From personal to digital: CCTV, the Panopticon, and the technological mediation of suspicion and social control', in D. Lyon (ed) *Surveillance as Social Sorting: Privacy, risk and digital discrimination*, London: Routledge, 249–281.

Norris, C. and Armstrong, G. (1999) *The Maximum Surveillance Society: The rise of CCTV*, Oxford: Berg Publishers.

Oc, T. and Tiesdell, S. (2000) 'Urban design approaches to safer city centres: The fortress, the panoptic, the regulatory and the animated', in J.R. Gold and G. Revill (eds) *Landscapes of Defence*, Harlow: Pearson Education, 188–280.

Pecora, V.P. (2002) 'The culture of surveillance', *Qualitative Sociology*, 25: 3, 345–358.

Pinck, P. (2000) 'From sofa to the crime scene: Skycam, local news and the televisual city', in M. Balshaw and L. Kennedy (eds) *Urban Space and Representation*, London: Pluto Press, 55–68.

Presdee, M. (2000) *Cultural Criminology and the Carnival of Crime*, London: Routledge.

Rivière, C. (2005) 'Mobile camera phones: a new form of "being together" in daily interpersonal communication', in R. Ling and P.E. Pedersen (eds) *Mobile Communications: Re-negotiation of the social sphere*, Surrey: Springer, 167–186.

Rose, G. (2004) ' "Everyone's cuddled up and it just looks really nice": An emotional geography of some mums and their family photos', *Social and Cultural Geography*, 5(4): 549–564.

Schienke, E.W. and IAA (2002) 'On the outside looking out: An interview with the Institute for Applied Autonomy (IAA)', *Surveillance and Society*, 1: 102–119.

Schmidt-Burkhardt, A. (2002) 'The all-seer: God's eye as proto-surveillance', in T.Y. Levin, U. Frohne and P. Weibel (eds) *Rhetorics of Surveillance from Bentham to Big Brother*, ZKM Centre for Art and Media: Karlsruhe, 17–31.

Staples, W.G. (2000) *Everyday Surveillance. Vigilance and visibility in postmodern life*. Lanham: Rowman & Littlefield.

Tabor, P. (2001) 'I am a videocam', in I. Borden, J. Kerr, J. Rendell and A. Pivaro (eds) *The Unknown City: Contesting architecture and social space*, Cambridge, Massachusetts: The MIT Press, 122–137.

Tinic, S. (2006) '(En)visioning the televisual audience: Revisiting questions of power in the age of interactive television', in K.D. Haggerty and R.V. Ericson (eds) *The New Politics of Surveillance and Visibility*, Toronto: University of Toronto Press, 308–326.

Weibel, P. (2002) 'Pleasure and the panoptic principle', in T.Y. Levin, U. Frohne and P. Weibel (eds) *Rhetorics of Surveillance from Bentham to Big Brother*, Karlsruhe: ZKM Centre for Art and Media, 207–223.

Whitaker, R. (1999) *The End of Privacy: How total surveillance is becoming a reality*, New York: The New Press.

Wise, J.M. (2004) 'An immense and unexpected field of action: Webcams, surveillance and everyday life', *Cultural Studies*, 18(2/3): 424–442.

Zizek, S. (2002) 'Big Brother, or, the triumph of the gaze over the eye', in T.Y. Levin, U. Frohne and P. Weibel (eds) *Rhetorics of Surveillance from Bentham to Big Brother*, Karlsruhe: ZKM Centre for Art and Media, 224–227.

Part IV

(In)secure virtualities

Chapter 8

The role of the Internet in the twenty-first-century prison

Insecure technologies in secure spaces

Yvonne Jewkes

Security fears as prisoners told: you have e-mail (Times, 2 January 2007)

Unlike many of the contributions in this volume, which discuss secure technologies in insecure spaces, this chapter is concerned with *in*secure technologies in secure places – specifically, internet use by inmates in prison. Apart from this paradigmatic relationship with the rest of the volume, the discussion that follows shares with other contributions several broad themes. First it is concerned with the interface between rights and security, and the curtailment of freedom and civil liberties of an already marginalised group. It explores these issues in the context of current access to the internet for prisoners in England and Wales, and argues that, in the twenty-first century, the denial of access to the internet constitutes a distinctive pain of imprisonment. Second, the chapter discusses how social relations come to be marked by distrust and suspicion, particularly with regard to people (in this case, convicted offenders) and technologies (the internet) defined as security threats. Third, the chapter notes that justifications made for denying prisoners access to the internet on grounds of security may be underpinned by more emotive objections based on nineteenth-century notions of 'less eligibility' and solitary confinement. Finally, in common with other contributions to this book, the chapter reflects on how the pursuit of security can result in greater levels of perceived and real insecurity, as policy on internet access in prison creates new forms of social exclusion.

A BRIEF SUMMARY OF THE CURRENT SITUATION

As anyone who occasionally reads a popular newspaper in the UK will be aware, mainstream press reporting of prisons tends to portray them as 'holiday camps' in which inmates enjoy luxuries they do not 'deserve'. Stories of this kind fuel the popular media's view of a criminal justice system which is soft on crime and which prioritises the requirements of offenders over those of victims.[1] Although these narratives might be dismissed as a trivialisation of prison issues, they serve to further stigmatise a population which is already at the margins and which rarely has a right of reply. Such stories also require escalating levels of hysteria and incredulity to make them newsworthy and so it is that after years of increasingly strident expressions of outrage about the cost and quality of prison meals, the provision of multi-faith places of worship, the introduction of personal television sets and various other 'perks', the notion of providing prisoners with access to e-mail and the internet has become a highly sensitive and controversial issue for any Minister of Justice to contemplate introducing.

Yet despite the considerable opposition voiced in the popular press and on internet forums, there have been significant moves forward in the last few years and several important initiatives are taking place across the prison estate. For example, following unsuccessful tests on software that restricts users to particular websites and disables the command key on their systems, at the time of writing trials were taking place at HMP Wandsworth in South London which not only permit access to the net but also to e-mail. In the realm of education and training, the All-Party Parliamentary Group for Further Education and Lifelong Learning has argued that facilities for distance learning and e-learning should be enhanced in every prison and supervised internet access made available to prisoners doing courses that require it. Some individual prison governors are known to be favourable to the idea of online learning: for example, the governor at HMP Chelmsford has pioneered a scheme whereby foreign national prisoners can study online in their own language on condition that they study English Language classes as well. Online learning organisation Learndirect, which operates a network of more than 2,000 e-learning centres, has installed servers and networked PCs in 20 prisons. These are used to deliver courses in literacy and numeracy and to impart skills for employment. Networking company Cisco has set up centres in 18 prisons under a scheme called the Prisons ICT Academy (PICTA). Several hundred prisoners have completed courses that cover basic computer skills and PC maintenance (http://www.literacytrust.org.uk/Database/prisonupdate.html). Another development is Summit Media, a digital media company which runs its operation from within HMP Wolds and HMP Rye Hill. At HMP Wolds, 25 prisoners have completed a full training programme in order to work producing websites and online marketing services to companies doing business on the web (http://www.hmpwolds.co.uk/main_pages/prison_industry.htm).

1 Although such stories are not confined to the 'red tops' but appear in the 'quality' press as well, the *Guardian* and *Observer* provide notable exceptions to the general trend with in-depth reporting of numerous contentious issues and problems facing prisoners, notably overcrowding, drugs, mental illness and deaths in custody. See http://www.guardian.co.uk/society/prisonsandprobation

While these initiatives are very important, they remain limited given the potential scope for prison internet access. Currently only seven prisons in the UK offer internet use, and it is exclusively for education and training. Further, in line with the more general education policy, the e-learning facilities and training that currently exist in prisons are directed at basic-level skills and are confined to supervised use of computer software rather than access to the world wide web. Vocational training is clearly important for the majority of prisoners who need skills as well as support to help them to resettle on release. But at the other end of the education spectrum, learning (as opposed to training), particularly in relation to degree programmes, is at risk of being squeezed. The government's policy of focusing almost exclusively on Level 1 literacy and numeracy is restrictive, especially for long-term and life-sentence inmates who frequently come to prison with high levels of education and skills.[2] The Open University has been the main provider of degree-level courses in prisons for the last 30 years and typically recruits over 300 students annually. However, in the last few years, the University has moved to online delivery of its courses, and the demands of prison security make online learning fraught with difficulty. Not only does the lack of internet access preclude degree-level study, but many prisoners are not allowed to possess CD-Roms or DVDs because the discs are considered potential weapons for assault or self-harm. Consequently, they have to make do with simulated tutorials that are loaded onto their computers rather than the real thing.

Past trials of e-mail exchange have been similarly compromised. In 2006 a six-month trial took place in HMYOI Aylesbury, HMP Downview and HMP High Down, which enabled prisoners' families and friends to write to prisoners via e-mail. The scheme, called PRIS-M, meant that e-mails were downloaded by prison mail room staff from a secure site on the internet and sent directly to a printer which automatically sealed each communication for privacy before being delivered to the prison wings. While prisoners were not permitted to send e-mails back, their handwritten letters were scanned onto a computer, uploaded to the PRIS-M server by mail room staff, and sent out electronically at a cost of 20 pence each.

2 While the low levels of literacy and numeracy that blight the prison population are inarguably a source of shame to UK society, when broken down into different kinds of establishments a more nuanced picture emerges. For example, according to Home Office evidence, 37 per cent of female prisoners have participated in further education, while research indicates that at one category 'D' prison in England, 69 per cent of prisoners had achieved GCSE 'O' Level grades or above prior to imprisonment; 29 per cent had 'A' Levels, and 31 per cent had a degree and/ or postgraduate qualifications (Hayward 2006).

A DISTINCTIVE PAIN OF (LATE) MODERN IMPRISONMENT

It is not difficult to imagine how great the impact of 'new' media technologies such as the internet could be on the lived experience of imprisonment. Like television and radio the internet 'creates new possibilities of being: of being in two places at once, or two times at once' (Scannell 1996: 91). But unlike traditional media, the internet allows us to interact with others anywhere in the world, in real time, and on equal terms. It is a 'many-to-many' medium, whereby everyone who is online is 'in the same place' (Holderness 1998: 35). Physical location and all the usual markers of identity are irrelevant. We can be anonymous, or invent an entirely new personality, or divulge aspects of our identity that would normally be kept hidden; we can 'connect' with like-minded individuals and groups, forming communities and alliances with others on the basis of shared interests rather than geographical proximity; we can enter worlds previously unknown to us and partake in events and experiences in contexts far removed from our own (Slevin 2000). More than any other medium, computer-mediated communications undermine the traditional relationship between physical context and social situation. Place and time are transcended. When we sit down at our computers and sign on to the internet we are no longer 'in' our physical setting but are relocated to a 'generalised elsewhere' of distant places and 'non-local' people (Morley and Robins 1995: 132; see also Meyrowitz 1985).

However, while the technological revolution that has occurred over the last two decades has expanded the social worlds of free citizens almost to the four corners of the globe, it has simultaneously created a new level of disconnection between prison and society (Johnson 2005). While most of us have seen our lives dramatically transformed by the convergence of internet and other user-driven, audio-visual technologies, including mobile phones, laptops, BlackBerrys and MP3s, prison inmates are limited to the most modern technology readily at their disposal – terrestrial television. Lack of access to information and communication technologies has thus been described as a 'distinctive pain of modern imprisonment' (Johnson 2005: 263).

Originating in the work of Gresham Sykes (1958), the 'pains of imprisonment' is a term used widely in prison sociology to denote the deprivations inherent in the experience of confinement that inflict particular distress and pain. Sykes identified five such 'pains': 1) the deprivation of liberty; 2) the deprivation of goods and services; 3) the deprivation of heterosexual relationships; 4) the deprivation of autonomy; and 5) the deprivation of security. Research has shown that access to media technologies in prison can help to ameliorate all these deprivations (Jewkes 2002). Loss of liberty is alleviated by the 'diversion' that media facilitate, in terms of both retreat from the immediate environment and emotional release. In an environment where everyday life is sometimes described in terms of its 'thinness', access to the internet would provide a richness, colour and texture which are, in some way, comparable to life outside. More than any other medium, the internet facilitates personal relationships with friends and strangers, counteracting feelings of loneliness and isolation. In addition, it provides a social utility function, giving people a source of conversation with others in 'real' time. Loss of liberty is also countered by the opportunities for surveillance of the world outside which the internet makes possible. The deprivation of goods and services that those on the outside take entirely for granted is easily understood, as is the desire to reach out to wider forms of community and alliance as a way of legitimising one's identity as, for example, a (potential) partner or parent. To be able to do so also improves a prisoner's sense of autonomy and goes some way to reducing the feeling of infantilisation engendered in prison. Finally, the heightened awareness of potential risks and insecurities that constitutes the paramount reality for many prisoners can be ameliorated by the construction of alternative life-worlds which privilege other emotional qualities – intimacy, companionship, humour, learning, relaxation, competitiveness and so forth. Alternative life-worlds are intrinsic to computer technologies and in prison, perhaps even more than in general life, media provide a refuge from the demands of public presentation and the rigours of social interaction.

While Sykes' work continues to be the touchstone for prison researchers who adopt a phenomenological approach, more recently scholars in the field have updated and developed his theory of prison 'deprivations' or 'pains' to make them more relevant to the modern-day custodial process. Among the most pertinent in the current context are: 1) loss of stimulation; 2); loss of social support; and 3) loss of communication (Jones 2007). Loss of stimulation means that boredom features heavily in a prisoner's life. In comparison with the mid 1990s, prisoners are spending more time locked in their cells and less time engaged in constructive or purposeful activities. Loss of stimulation may be particularly difficult for young people to cope with: it is estimated, for example, that at Glen Parva YOI, 30 per cent of prisoners can be locked in their cells at any given time (HMIP 2004). Furthermore, the Social Exclusion Unit (2002) found that in 2000, young adult prisoners aged 18–20 spent only 23.1 hours per week on purposeful activity. In short, a great deal of prison life is spent doing nothing.

Extending from Sykes' deprivation of heterosexual relationships is a loss of social support (Jones 2007). Separation from family and friends is one of the greatest pains faced by most prisoners, and is felt especially acutely by young prisoners for whom family is a vital form of support. For example, Harvey (2007) found that for young men in prison, separation from loved ones is their biggest concern and that the most important ties are with their mothers and relationship partners. In this respect, young male prisoners occupy a vulnerable position; many have strong ties with their own parents, yet are simultaneously fathers to their own children. Loss of social support can have a devastating effect on such individuals and is a high-risk factor in suicidal behaviour and self-harm (Jones 2007).

Relatedly, the restricted opportunity to communicate with those on the outside has been identified as a significant pain of imprisonment (Jones 2007). Many prisoners are accommodated a long way from their family homes and consequently receive few or no visits. The finding that British Telecom charges prisoners more than five times the standard payphone rate to use a prison phone may explain why many prisoners are discouraged from maintaining family ties (Allison 2006). Letter writing frequently involves delays and does not come easily to all prisoners; in fact, 80 per cent of prisoners have writing skills at or below the level of an 11-year-old child (Social Exclusion Unit 2002). Among the most disadvantaged in this respect are offenders aged between 18 and 20.

Giving prisoners access to the internet and e-mail would diminish all these problems and allow them a form of communication which, unlike letter writing, is instantaneous, interactive and part of most young people's everyday lives. Research shows that in the wider community, 91 per cent of 16–21 year olds not in education have access to mobile phones, the internet and e-mail, with 55 per cent of 11–21-year-old males feeling unable to be without video games on Play Stations or personal computers (Haste 2005). Furthermore, 97 per cent have access to a computer that links to the internet (Haste 2005). Young people have come to expect almost constant stimulation and communication, and in prison, permission to write letters and use a limited number of shared payphones simply does not compensate for losing access to mobile phones and e-mail (Jones 2007). Use of the internet would also permit parents in prison to stay in touch with their children via e-mail or social networking sites such as MySpace and Facebook, and would give children and young people in custody a familiar cyber space.

Unsurprisingly, many prisoners believe that the restricted access they have to new communication technologies and, in particular the almost total absence of computers and internet access, is a form of censure that renders them second-class citizens in the Information Age. Far from sharing with the wider society the privileges of the advancing communications networks, prisoners are impoverished by their lack of technological hardware and by their concomitant inability to exchange information in ways that have become commonplace for most of us. Prisoners are also largely immune from the transformations of time and space that have arisen from media technologies. While most of us are aculturised to a world where time is speeded up, slowed down, suspended, repackaged, re-ordered and re-experienced through digital and satellite technologies, most prison inmates experience time in a more traditional, chronological sense and exist through time in a much more linear fashion, almost as if in a pre-media age. These obstacles arguably render prisoners, especially those serving long sentences, 'cavemen in an era of speed-of-light technology' (Johnson 2005: 263).

THE FRAMING OF PRISON INTERNET AS A QUESTION OF SECURITY

The reasons why politicians and policymakers have been slow to acknowledge the benefits to prisoners of giving them relatively unrestricted internet access may be complex, but the justification most frequently voiced is quite straightforward. Issues of security have come to shape and dominate debates about whether prisoners should be allowed to use interactive computer technologies. As we shall see later in the chapter, attempts to thwart internet access in fact may be underpinned by entirely different motives and considerations. But political justifications are founded on perceptions of the technology's inherent insecurity; a rationalisation that is difficult to counter within a system of governance characterised by audit, accountability and assessments of risk (Feeley and Simon 1992). Specifically, official resistance to prison internet access centres primarily on the possibility that it will be used by prisoners to view pornography, contact victims, intimidate witnesses and plot escapes.

To some extent, such fears have previously been rehearsed in relation to other communication technologies, but have been either successfully managed or dismissed. For example, following the introduction of telephones for prisoners' use in 1988, concerns were expressed that prisoners could contact witnesses and/or victims of their offence. The phonecards that were issued were subsequently replaced by PIN phones, whereby a list of the telephone numbers prisoners wish to call is submitted for approval and the cost of calls is deducted from credit in their PIN phone account (Mills 2008). To take another example, when in-cell television was first mooted, anxieties were voiced because it is difficult and impractical to try to censor TV content on personal sets. Concern focused on the possibilities that individuals sentenced for violent and/or sexual offences would be able to view violent and/or sexual material, and that TV images might reinforce, or even legitimate, prisoners' criminal identities (Jewkes 2002). Certainly there is evidence that some prisoners watch programmes that use covert filming techniques to see whether they know the offenders caught on camera, while others view TV programmes to learn 'new tricks' and improve their criminal expertise, or to pick up legal knowledge to help prepare defence cases or campaign for the political and social rights of prisoners (Hendrick 1977; Hagell and Newburn 1994; Jewkes 2002). Nevertheless, most prison governors view its potential drawbacks as relatively insignificant within the broader picture of prisoners' rights to enjoy the same entertainment as the rest of society, and in-cell television was rolled-out across the prison estate in the late 1990s. The interactive nature of the internet makes it a very different proposition however.

The merging of fears about offenders with fears about technology

In recent years a number of high-profile, salaciously reported internet offences have come to public attention, leading to calls for greater self-regulation, tougher legislation and even censorship. Anxiety about the power of the internet to influence dangerous or vulnerable users reached an apotheosis when the headline 'Killed by the internet' appeared in the *Daily Mirror*, a British tabloid, on 5 February 2004 (Jewkes 2007). Since that time, reports have circulated about dozens of serious assaults, abductions and deaths of individuals in countries around the world that are said to be internet assisted, including, in November 2007, the story of so-called 'YouTube killer' Pekka-Eric Auvinen, a Finnish student who shot dead eight people at his high school in an incident reported as 'spurred by the internet and the isolation of a troubled teenager' (*Times*, 8 November 2007). Not only do the media over-report atypical crimes, they under-report the experiences of marginalisation and social exclusion that offenders commonly experience throughout their lives. In the current penal climate, politicians and the media discuss individual moral responsibility as if it exists in a vacuum, somehow detached from the circumstances in which people find themselves (Drakeford and Vanstone 1996). As Mythen and Walklate observe, this leaves little room for rational attempts to understand the values, objectives and/or grievances of these individuals and instead reduces the offender to 'an inhuman object of hate' (2006: 10).

Of course, there is nothing inherently sinister in the technology itself. Most cyber-crimes are reasonably common offences; computer technologies have simply provided a new means to commit 'old' crimes, and it is clearly not the case that if the internet did not exist, neither would violent and sexual crimes. What makes the role of the internet unique are its interactive capacity, its sheer scale and reach, and the way that it has opened up new channels for the presentation of self. As users, we can be anonymous or we can expose our inner psyches to a global audience. Not only does the net provide us with the opportunity to present, or hide, aspects of our selves, it also permits us to invent new selves. In the virtual world, identity is multi-dimensional and amorphous; we can be whoever, whatever, wherever we wish to be. The internet is the slate upon which we can write and re-write our personalities in a perpetual act of self-creation. Considering that many cyber-crimes are 'underground' activities carried out in 'clubby' atmospheres in the company of like-minded individuals, and that they carry a relatively low risk of detection, it is little wonder that the internet has become a scapegoat for a series of local and global moral panics.

Sensationalised media reports of atypical cases reinforce notions of technological determinism and encourage circumscribed and predictable responses. At its mildest, opposition to the inclusion of prisoners in the mediated public sphere is founded on the fact that those 'on the outside' can no longer use media technologies as a private forum in which to discuss the problems of crime, crime prevention, punishment and so forth, since inmates can now 'enter' society via media access. Further, as prisoners are increasingly able to monitor and interact with the larger environment informationally, it is feared that they will correspondingly increase their demands for greater physical access to the outside world and expect entitlements commensurate with those accorded the wider population. These two processes arguably create a shift in the balance of power, so that instead of normalisation happening at the pace at which the prison service thinks appropriate, prisoners are themselves playing a role in change (Meyrowitz 1985). In more extreme manifestations, there is a constant drip of frankly astonishing stories about the internet's potential to corrupt, fed to us by a popular media baying for tougher laws to deal with cyber-offenders. Like other 'new' technologies, which are frequently referred to as Janus-faced because they appear to offer palpable benefits while, at the same time, creating new risks and uncertainties (Lyon 1994, 2001), the internet elicits excitement and fear in equal measure. Thus, despite transforming most people's lives for the better, the internet has, at the same time, become the repository for many of society's darkest fears, and notions of cyber-space as a lawless jungle or postmodern version of the Wild West prevail.

That debates about this subject have merged security fears about offenders with anxieties about the nature of the technology underlines the difficulties faced by those who advocate prison internet access. The fact that the internet allows users to conceal undesirable or stigmatised aspects of their 'real' identities while simultaneously adopting new identities, and that it creates and supports amorphous 'communities' which are not only are anonymous but in constant states of flux and impermanence, is anathema to a society that seeks to identify, label and contain its offenders. As a result, instead of putting resources into the conditions under which internet access in prisons *can* be administered and governed, and emphasising the positive outcomes of a technology that could help the Prison Service to manage dispersed relationships, policy questions have been situated within an inarguable logic of security. It is simply one example of the widespread political technique of framing policy questions in the logics of 'common sense'; a practice that works to characterise social relations on the basis of suspicion and distrust (cf. Huysmans 2006).

It has been suggested that the mobilisation of a politics of fear dangerously amalgamates authoritarian populism with pseudo-scientific critique, resulting in a form of 'risk-crazed governance' (Carlen 2008). Many scholars have argued that there exists in late modern society a pervasive social risk-consciousness, which is nurtured and utilised in penal settings as a tool of control (see, for example, Garland 2002; Carlen 2008). Against this background, insecurity becomes a domain of practice as opposed to a definition of threat (Huysmans 2006) and any suggestion of prisoners having access to the internet is conveyed to the wider population in tones of outrage and derision. In the absence of alternative opinions, such as those that might be expressed by security experts or prison educators, media audiences are encouraged by politicians and policymakers to take up a particular stance in relation to the issue. Those politicians and policymakers then respond punitively, claiming to speak and act on behalf of the people, and, in turn, punitive penal policies are reported back to the people by the media. This cyclical process is highly successful in shifting the ordinary – or, in the case of the internet, the *extra*ordinary – into the threatening.

Of course, the fact that 'public opinion' is usually articulated and filtered by the popular press renders statements based on public sentiment entirely unreliable, and the assumption that the public are uniformly and unremittingly hostile to prisoners being granted access to the internet may be considerably overstated. While the public mood can be judged to darken at times when a particularly severe violation of security occurs within prison walls, it would be misleading to suggest that opinion is homogenised on prison matters, or that decisions such as the introduction of internet technologies in prisons are simply fought out between two monolithic and oppositional forces: the 'penal populist brigade' versus the 'liberal do-gooders'. In truth, most people go through their lives having no direct contact with a prison or prisoners and are quite ignorant of the minutiae of everyday life for those confined. Fed a diet of media stories about notorious offenders, violent assaults and security breaches, little thought may be given to the things that normalise prison regimes such as access to education, decent healthcare, or providing a means of communication between inmates and their families. However, when sensational sound-bites are replaced by thoughtful and informed debate, public opinion may not be so predictable. Illustrating the complexities of the issue, a BBC Radio 4 discussion of the subject, led by presenter Libby Purves, introduced the subject of prisoners having access to the internet as an 'alarming idea' to many people. Yet a poll conducted during the course of the programme, which included two 'experts' who were pro-internet access in prisons, found that 83 per cent of listeners were in favour of it (The Learning Curve, BBC Radio 4, 24 May 2005; see http://www.bbc.co.uk/radio4/factual/learningcurve_20050524.shtml).

Nevertheless, the notion that it is public opinion that drives decision making in this area remains in common currency and is enshrined in policy. For example, in England and Wales media use has always been integrated into the system of Incentives and Earned Privileges (IEP), and one of the aims of IEP is to meet 'public expectations about what kind of place prison should be' (Liebling et al 1997). Former prisoner and *Guardian* writer Erwin James illustrates the outcome of this policy:

> A few months after I started writing my column, 'A Life Inside', for the

Guardian some seven years ago, I asked the governor of the prison I was in if I could have a word processor sent in by supportive friends.

'Oh no,' he said after a sharp intake of breath. 'The public wouldn't like that.'

I wasn't sure how exactly they would find out, and if they did I could always hope that there were some who would see it as an aid to a successful reintegration into society if I ever were to be released.

'But word processing is a transferable work skill,' I said, 'essential for a (ahem) writer.' He still wasn't convinced.

'We've got to think about the climate out there,' he said pointing towards the local townsfolk. And that was the end of the matter.

(*Guardian*, 17 September 2007)

TWENTY-FIRST CENTURY VERSIONS OF SEPARATION, SILENCE AND LESS ELIGIBILITY

The governor's response to James quoted above subtly demonstrates how populism has become a marked feature of penal politics in the last couple of decades (Pratt 2007). Although populist arguments about security make policies of denial (e.g. of prison internet access) intelligible, the language of penal populism also suggests that the conflation of distinct sources of anxiety under the name of 'security' may actually be underpinned by more emotive responses. Where offenders are viewed as 'more numerous, more threatening, more undeserving, less corrigible and, perhaps, less akin to ourselves then priorities accordingly tend to refocus on deterrence and secure confinement' (Sparks 2007: 91).

It hardly needs stating that notions of prisoner empowerment and freedom do not sit easily with modern political rhetoric, which – as Erwin James has reminded us – is more concerned with satisfying perceived public demands than with prisoners' rights. But if, as the adage goes, individuals are sent to prison *as* punishment, not *for* punishment, why is it that control over communication between prisoners and other prisoners, and between prisoners and the outside world, remains central to current ideas about punishment? The historical antecedents for prisoners' social isolation are well documented (see for example Henriques 1972; Forsythe 1987; McGowen 1998). Since the early nineteenth century, prisons have been more than places of physical incarceration; they have been places of informational isolation as well. The two disciplinary regimes that dominated nineteenth-century prisons – the separate and silent systems – denied prisoners contact and communication with others. Limited in movement, isolated from other prisoners, forbidden from making even the smallest of gestures to a fellow inmate and effectively ex-communicated from society, the Victorian prisoner was subjected to a brutal and solitary regime which was nonetheless regarded as being 'too soft' by sections of the media of the day (Johnston 2006).

One hundred and fifty years on, the prison system allows most prisoners to share with the wider society the 'privileges' of radio, television and telephone, which gives governors and other prison staff a raft of effective penalties for relatively minor transgressions including withholding visits, prohibiting prisoners from making telephone calls and confiscating in-cell television sets. The 'carrot-and-stick' mentality that underlies most forms of communication within prison causes many prisoners to be ambivalent about them, and there is widespread resentment among the inmate population that technology in prisons tends to be used for purposes of control and punishment rather than reform or rehabilitation (Jewkes 2002; Johnson 2005). The control and curtailment of interpersonal communication also underpin solitary confinement; usually regarded as the severest form of non-capital punishment in the West. Exemplified in the United States by the 'supermax', prisoners at some of these super-maximum security institutions are completely isolated not just from fellow prisoners but also from staff, who remain behind physical barriers. In these establishments, prisoners do everything in isolation, including exercising in 'dog runs' (King 2008).

In addition to severely restricting prisoners' contact and communication with others, Victorian prisons adhered to the principle of 'less eligibility' – a doctrine of deterrence demanding that prisoners should endure material living conditions which compare unfavourably to those of similarly disadvantaged yet 'decent' people in the community. The legacy of less eligibility is evident in the current language of penal austerity, which recalls and reiterates nineteenth-century ideas about the 'deserving' and 'undeserving' poor (Sparks 1996). Since that time, the characterisation of prisoners as second-class citizens who have forfeited their rights to any comforts or privileges has remained prevalent in our culture; hence reports concerning anything from prisoners' Christmas dinner to young offenders being allowed to play video games are recounted as if gruel and treadmills would be more fitting. The principle of less eligibility even extends to education and training, despite the rehabilitative benefits they promise. For example, when it was revealed in 2001 that the boys convicted of killing James Bulger in 1993 had attained 'A' Levels and received education and training opportunities that they would not have enjoyed had they not been detained in custody, the English media responded with anger and incredulity.

Similar expressions of outrage, amalgamating repressive attitudes to prisoners' freedom to communicate with autocratic notions of less eligibility, have accompanied the introduction of 'new' media technologies into prisons. Take this erroneous report in the *Daily Mirror*:

Power perks for jail's dangermen

Prisoners at a top security jail are to be allowed their own hi-tech gadgets, including laptop computers, CD players and computer games.

The move, aimed at stopping inmates tapping into the electricity supply, follows the wiring up of 100 cells. But jail officers at Frankland Prison, Co Durham, say it is another sign of the regime bowing to pressure from prisoners and could turn the place into a 'hotel' . . .

Earlier this week, prison chiefs replaced new seats in the visiting area after convicts complained they were uncomfortable and stopped them touching their loved ones.

(*Daily Mirror*, 6 May 1995)

More than a decade later, the popular press still report media technologies as luxuries that prisoners do not deserve:

Cons TV enough to make you sick

It's cheaper than hospital

Fury erupted last night as it emerged prisoners can watch TV in their cells for £1 A WEEK – but NHS patients have to pay £3.50 A DAY.

Some jails, such as Saughton in Edinburgh, even offer satellite TV packages, including live football. . . .

MSP Michael Matheson stormed: 'It is simply outrageous that patients pay so much more than cons.'

(*The Scottish Sun*, 20 March 2006, cited in Sparks 2007)

These media reports underline the point made earlier that generalised claims are given authority and legitimacy by reference to public outrage and indignation, yet such opposition frequently tends to be supported by quotes from prison officers, politicians or newspaper editors expressing *their* hostility to the idea, rather than by members of the public themselves.

Two ironies inherent in the pursuit of security

That the prison/internet debate has been dominated by the twin languages of austerity and security has resulted in two profound ironies. The first is that, if and when prisoners are permitted routinely to access the net, the technology that promises so much freedom is likely to be implemented by the authorities in ways that enhance security and surveillance. Control takes many forms, including the seemingly benign, and computer technologies will undoubtedly follow in-cell TV as a means of managing prisoners' inclusion and exclusion both from certain spaces at certain times and from certain levels of 'privileges'. Johnson (2005) notes that, in the USA, prison visits, telephone calls, work release programmes, compassionate leave, permission to decorate cells and keep pets, facilities to cook one's own food and permission to receive personal property and wear civilian clothes have all been eroded in prisons where the one 'perk' allowed is access to television. A similar pattern of social and behavioural control is occurring in the UK where, perversely, the introduction of media resources into prisons may be reproducing disadvantage and deprivation. Earlier lock-up times have been introduced, education opportunities have been curtailed and spaces for inmates to interact with others have been restricted. It is difficult to avoid the conclusion that personal media have one great, unspoken advantage as far as prison authorities are concerned, and that is to normalise the regulation and surveillance of inmates (Jewkes 2002).

The second irony is that, for as long as prisoners' access to the internet is framed as a security issue, the repercussions are likely to involve greater *in*security for the community at large, as prisoners are released back into the community with significant skills deficits. Research on desistance and recidivism shows that two of the most important factors in determining offenders' desistance from crime when they leave prison are lack of employment opportunities and loss of accommodation. In practical terms, access to computers, e-mail and the internet would allow users in prison to interact with potential employers and teach them the information technology skills that many jobs now require. Prisoners could also prepare for release by being able to contact public-sector organisations that offer information on their websites about issues related to housing and hostels.

A further obstacle to the process of 'going straight' is loss of contact with families and, in particular, damaged or severed relationships with children. As we have already seen, computer-mediated communications could be an immensely valuable tool for the Prison Service in allowing prisoners to sustain relationships with family and friends, as well as tutors and lawyers. As it is, the disruption of family relationships and employment opportunities that even short periods of confinement can entail creates feelings of being 'held back' and can act as a breeding ground for future criminality (Farrall 2008).

In addition, internet access would provide prisons with a wider range of resources for delivering effective courses, and offer prisoners and staff opportunities for the acquisition of new skills. Over half of all male prisoners have no qualifications at all and the failure of other agencies to deal with these social problems leaves the Prison Service and its partners with the task of 'putting right' a lifetime of service failure. Consequently, along with strategies to address offending behaviour and reconviction rates, the teaching of basic skills has become a priority in prisons, not least because the three are believed to be connected. However, only about a third of prisoners are offered access to education, and it is not compulsory for adult prisoners. Many prisoners who are offered education decline it because they get paid more for doing menial work around the prison, such as cleaning the wings or picking up litter from the grounds. Further, the government's instrumentalist approach to prisoner education and the formal linking of basic skills to a reduction in recidivism in key performance indicators is of concern to many who believe that *lifelong* education slows the revolving door of incarceration and re-incarceration (Hayward 2006). Denying prisoners internet access might thus be seen as an example of technology being used as a strategy of social exclusion.

It is entirely plausible, then, that the consequences of not allowing prisoners access to online education, information and entertainment are considerably greater than the potential security threats posed by any individual inmate. Certainly, a technologically illiterate prisoner population cannot be regarded as desirable in a fast-moving and technologically advanced society. That this truth tends to be subsumed by the overriding view that the more humane prison regimes become, the less effective they are as a deterrent, obscures the real paradox of prisons, which is that they are places of human(e) aspiration and well-meaning social experimentation, as well as sites of struggle, abuse and neglect (Sparks 2007). Granted, in a political climate where ministers clamour to 'out-tough' each other on law and order matters, politicians and policymakers arguably have greater fear of embarrassment if revelations appear in newspapers about prisoners having computers than if disclosures are made about prisoners locked up for 23 hours a day in an overcrowded prison (Stern 1987). It is also true to say that the criminal justice system has become a primary arena for politicians and policy makers to display their macho credentials, and it is widely accepted that prisoners' rights is not a vote-winning issue. However, most interested parties (e.g. the now defunct Forum on Prisoner Education and Pipeline, a pan-European project set up to share information and good practice on prisons and the internet) remain optimistic that introduction of the internet in prisons *will* happen, although in the UK change may occur only when the Prison Service is forced, through, for example, human rights legislation, to permit inmates to use computers.[3]

CONCLUDING THOUGHTS

As we have seen, in their ability to liberate users from the usual constraints of corporeality, and to bring the outside inside, the potential benefits of

3 This is already happening to a degree: there have been a small number of successful legal challenges by prisoners who have won the right to use laptop computers to prepare a defence.

computer-mediated communications to prisoners are incalculable. In practical terms, access to computers, e-mail and the internet would allow users in prison to interact with potential employers and public-sector organisations that might help with particular issues such as housing prior to release, and increase contact with tutors, lawyers and family. The internet could be an immensely valuable tool for the Prison Service in handling fragmented and fragile relationships and could also provide prisons with a wider range of resources for delivering effective courses, offering both prisoners and staff opportunities for the acquisition of new skills.

However, it is not difficult to conceive of why late-modern democratic governments perpetuate the politics of insecurity. In contemporary society, shared fears constitute community and give a powerful mandate to increasingly repressive forms of governance. As other chapters in this collection illustrate, there exist numerous examples of political and socially constructed manifestations of security and insecurity designed to mobilise public anxieties, broaden the scope of state powers and legitimise discriminatory policies. In the realm of crime and justice, politicians and the media have jumped on the populist bandwagon, perpetuating the notion that people commit offences because 'they' are not like 'us'. Further, as Carlen (2008) notes, when civil liberties or human rights are threatened by policy, then evidence, ethics and logic are abandoned and the public are expected to trust politicians' judgements. On the issue of prisoners' rights to access the internet, 'knowledge' is being compartmentalised and essentialised through selective promotion of those opinions supportive of official depictions of risk and the necessary strategies for its governance. Meanwhile other voices are silenced, e.g. those of security experts with the knowledge and skills to suggest how internet use in prisons could best be managed, and alternative constructions of internet access – which might include a humanitarian approach underpinned by the belief that prisoners possess human rights that must be respected, or, alternatively, that it is a profoundly important matter of rehabilitation and resettlement – are absent from mainstream political and media discourse (cf. Mathiesen 2004; Huysmans 2006).

Underpinning political justifications about security are atavistic notions of less eligibility, which are arguably the more salient reason why progress of internet access in prisons has been slow. The idea of prisoners using the internet to communicate, learn, play games and shop like the rest of us do may fuel resentment, not least among those who work within the popular media who perpetuate the notion that prison is a 'kind of country club for the lower classes' (Johnson 2005: 256). Viewed in this way, prisoners' access to the internet has become part of the landscape of 'imaginary penalities' that suppresses other forms of knowledge (Carlen 2008). The result is that both the debate about whether prisoners *should* have access to computer-mediated communications and the populist political denial of access serve to highlight offenders' 'otherness' and reinforce punitive approaches to criminal

governance. Access to computer-mediated communications has simply added the technological variables of 'high information' versus 'low information' prisons to the physical variables of 'high security' and 'low security' (cf. Meyrowitz 1985), but at the beginning of the twenty-first century, prisons are still falling back on nineteenth-century notions of punishment. Framing prisoners' rights to use the internet as a security matter makes the policy of denial intelligible, but it obscures the reality, which is that in contemporary penal philosophy, segregation, separation and silence remain the severest penalties.

References

Allison, E. (2006) 'Phone call costs cut off prisoners', *Guardian* (online). Available: http://society.guardian.co.uk/offdiary/story/0,,1683160,00.html

Carlen, P. (ed) (2008) *Imaginary Penalities*, Cullompton: Willan.

Drakeford, M. and Vanstone, M. (eds) (1996) *Beyond Offending Behaviour*, Aldershot: Ashgate.

Farrall, S. (2008) 'Desistance', in Y. Jewkes and J. Bennett (eds) *Dictionary of Prisons and Punishment*, Cullompton: Willan.

Feeley, M. and Simon, J. (1992) 'The new penology: Notes on the emerging strategy of corrections and its implications', *Criminology*, 30(4): 449–474.

Forsythe, W.J. (1987) *The Reform of Prisoners, 1830–1900*, Beckenham: Croom Helm.

Garland, D. (2002) *The Culture of Control: Crime and social order in contemporary society*, Oxford: Oxford University Press.

Hagell, A. and Newburn, T. (1994) *Young Offenders and the Media*, London: Policy Studies Institute.

Harvey, J. (2007) *Young Men in Prison: Surviving and adapting to life inside*, Cullompton: Willan.

Haste, H. (2005) *Joined-up Texting* (online). Available: www.mori.com/polls/2004/pdf/nestlesrp3.pdf

Hayward, D. (2006) 'Higher barriers: Ex-prisoners and university admissions', in S. Taylor (ed) *Prison(er) Education*, second edition, London: Forum on Prisoner Education.

Hendrick, G.H. (1977) 'When television is a school for criminals', *TV Guide*, 29 Jan.

Henriques, U.R.Q. (1972) 'The rise and decline of the separate system', *Past and Present*, 54: 61–93.

HM Inspectorate of Prisons (2004) *Prisoners Under Escort* (Online). Available at: http://www.inspectorates.homeoffice.gov.uk/hmiprisons/thematic-reports1/under-escort-04.pdf?view=Binary

Holderness, M. (1998) 'Who are the world's information-poor?', in B. Loader (ed) *Cyberspace Divide: Equality, agency and policy in the information society*, London: Routledge.

Huysmans, J. (2006) *The Politics of Insecurity: Fear, migration and asylum in the EU*, London: Routledge.

Jewkes, Y. (2007) *Crime Online*, Cullompton: Willan.

Jewkes, Y. (2002) *Captive Audience: Media, masculinity and power in prisons*, Cullompton: Willan.

Johnson, R. (2005) 'Brave new prisons: The growing social isolation of modern penal institutions', in A. Liebling and S. Maruna (eds) *The Effects of Imprisonment*, Cullompton: Willan.

Johnston, H. (2006) ' "Buried alive": Representations of the separate system in Victorian England', in P. Mason (ed) *Captured by the Media: Prison discourse in popular culture*, Cullompton: Willan.

Jones, H. (2007) 'The pains of custody: How young men cope through the criminal justice system', unpublished PhD thesis, University of Hull.

King, R. (2008) 'Supermax prisons', in Y. Jewkes and J. Bennett (eds) *Dictionary of Prisons and Punishment*, Cullompton: Willan.

Liebling, A., Muir, G., Rose, G. and Bottoms, A. (1997) *An Evaluation of Incentives and Earned Privileges: Final report to the Prison Service*, 1 July.

Lyon, D. (2001) *Surveillance Society: Monitoring everyday life*, Buckingham: Open University Press.

Lyon, D. (1994) *The Electronic Eye: The rise of surveillance society*, Cambridge: Polity.

Mathiesen, T. (2004) *Silently Silenced: Essays on the creation of acquiescence in modern society*, Winchester: Waterside.

McGowen, R. (1998) 'The well-ordered prison: England, 1780–1865', in N. Morris and D.J. Rothman (eds) *The Oxford History of the Prison: The practice of punishment in Western society*, Oxford: Oxford University Press.

Meyrowitz, J. (1985) *No Sense of Place: The impact of electronic media on social behaviour*, Oxford: Oxford University Press.

Mills, A. (2008) 'Communication', in Y. Jewkes and J. Bennett (eds) *Dictionary of Prisons and Punishment*, Cullompton: Willan.

Morley, D. and Robins, K. (1995) *Spaces of Identity: Global media, electronic landscapes and cultural boundaries*, London: Routledge.

Mythen, G. and Walklate, S. (2006) 'Communicating the terrorist risk: Harnessing a culture of fear?', *Crime, Media, Culture*, 2(2): 123–142.

Pratt, J. (2007) *Penal Populism*, Abingdon: Routledge.

Scannell, P. (1996) *Radio, Television and Modern Life*, Oxford: Blackwell.

Slevin, J. (2000) *The Internet and Society*, London: Routledge.

Social Exclusion Unit (2002) *Reducing Re-Offending by Ex-Prisoners*, London: Social Exclusion Unit.

Sparks, R. (2007) 'The politics of imprisonment', in Y. Jewkes (ed) *Handbook on Prisons*, Cullompton: Willan.

Sparks, R. (1996) 'Penal austerity: The doctrine of less eligibility reborn?', in R. Matthews and P. Francis (eds) *Prisons 2000*, London: Macmillan.

Stern, V. (1987) *Bricks of Shame: Britain's prisons*, Harmondsworth: Penguin.

Sykes, G. (1958) *The Society of Captives: A study of a maximum security prison*, New Jersey: Princeton University Press.

Chapter 9

Computer crime control as industry

Virtual insecurity and the market for private policing

Majid Yar

It has been argued recently that societal and systemic dispositions towards security are undergoing (or have already undergone) a fundamental shift. The central dimension of this transition can be traced in the reconfiguration of how we collectively conceive and manage security across a whole range of spheres (variously economic, social, political, environmental and interpersonal). In what Zedner (2007: 262) calls a 'pre-crime society', there is a shift in 'the temporal perspective to anticipate and forestall that which has not yet occurred and may never do so'. In the place of *ex-post* reaction to undesirable and harmful events, 'there is calculation, risk and uncertainty, surveillance, precaution, prudentialism, moral hazard, prevention and, arching over all these, there is the pursuit of security' (ibid.). This movement is manifested as the revelation of threats to security that *may* entail deleterious consequences and whose threat must be managed so as to curtail them before potentiality of harm becomes actuality. Thus we witness an incessant, nagging *insecurity* about the carcinogenic properties of mobile phones (Burgess 2004), an 'epidemic' of childhood obesity (Kline 2004), recreational drug use amongst young people (Manning 2007), MMR vaccinations (Stroud 2005), the threat of sexual assault via 'date rape' (Moore and Valverde 2000), and Islamic terrorism (Welch 2006; Furedi 2007), amongst many other issues. Such insecurities are no doubt incited by the drive to order, control and regulate a world that is perceived as ever more risky (symptomatic of what Furedi (2005, 2006) calls a 'culture of fear'; see also Glassner (2000)). Yet in turn they call forth a host of interventions that aim to 'securitise' what is perceived as a dangerously disordered, mutable and unpredictable social field. Thus insecurity generates security seeking, and security-seeking contributes to a greater sensitivity to risks, thereby contributing to greater feelings of insecurity (see also discussion in Ericson 2007). The apparatus of securitisation has crucially entailed the involvement of actors located outside the 'traditional' institutions of governance, namely those of the state. Instead, in decidedly neo-liberal times, the capitalist market now mediates a bewildering range and variety of tools and technologies, expert knowledge and 'best practice' strategies, precautionary codes and threat assessments.

Security has, in other words, now become inextricably entwined with the circuits of accumulation in contemporary capitalism.

The present chapter takes as its focus one such domain of insecurity and securitisation, that relating to computer crime and information security. The growing societal dependence upon and use of computerised systems, and not least the centrality of electronic communication networks such as the internet, has brought in its wake a host of (real or perceived) insecurities. Familiar instances include elevated concerns about internet child pornography and child exploitation, intellectual property theft and 'piracy', cyber-stalking, hate speech, identity theft and fraud. Each of these has stimulated an appetite for measures that could eradicate, curtail or manage the risks that they supposedly represent. Instead of reliance upon what Loader (2006) calls the 'platonic guardians' of public protection, there has emerged an extended market-led sector that sells security against the threats and predations that are seen to pervade contemporary cyber-worlds.

Significant attention has been devoted in recent criminological literature to the emergence of privatised forms of crime control in western industrialised societies. Amongst the empirical developments subjected to analysis have been the emergence of the commercial sector in prisons and corrections (see, inter alia, Ericson et al 1987; Feeley 1991; Christie 2000), the growth of private policing and security services (see, inter alia, Shearing and Stenning (eds) 1987; Johnston 1992; South 1994; Bayley and Shearing 1996), the commercial provision of products to the criminal justice system (Lilly and Knepper 1992) and the provision of security goods and services to commercial organisations, communities and individuals. However, one of the most recent developments of this 'crime control industry' (Christie 2000), namely the dramatic growth of computer and internet security and 'cyber-crime control', remains as yet largely unexplored by criminologists. This is all the more surprising given that this sector of commercial crime-control provision, usually identified as the 'computer security industry', is now worth an estimated $27 billion per annum (Grow 2004) and continues to grow apace.

In this chapter I chart the growth of this sector and attempt to situate its emergence in the wider contexts of social, political and economic change. I argue that the rapid upward trend in the commercial market for computer security can be best understood in relation to the conjunction of a range of processes. First, at the systemic level, I identify the move towards a regime of 'neo-liberal governance' in crime control, one that increasingly supplements the state-led, hierarchical provision of policing with a distributed, market-led network. Second, I note the consolidation of an 'information economy' in which networked computer technologies (such as the internet) are increasingly central to the development and delivery of goods and services, and in which intellectual property and its protection is seen as the central focus of a 'post-industrial' economy. Third, and most recently, I assess the impact of the September 11 attacks, which has elevated concerns about the threat of 'cyber-terrorism' and has served to further stimulate the demand for products and services that will protect critical national and commercial information infrastructures. The ongoing impact of these developments, I suggest, indicates that the computer security industry is likely to be at the forefront of the commercial provision of crime control in the early years of the twenty-first century. In the fourth and final section of the chapter I adapt the research programme propsed by Bayley and Shearing (2001) to map out an agenda for further inquiry into the growth, shape, drivers and regulatory and social implications of the computer crime-control industry.

THE COMPUTER CRIME-CONTROL INDUSTRY: SCOPE AND SCALE

The computer crime-control industry (CCCI) provides a wide range of IT security products and services on a commercial basis. These products and services are variously concerned with safeguarding the integrity and operation of computer systems, controlling access to systems and protecting the data content of systems from theft, unauthorised disclosure and alteration. The provision of such security measures takes an ever-expanding array of forms, including:

- security services, such as consultancy on threat analysis, systems security design and implementation, contingency planning, and disaster recovery (Nugent and Raisinghani 2002: 7);
- design and provision of software for user authentication and controlling access (e.g. password systems, smart cards and, most recently, biometrics such as fingerprinting and face recognition) (Halverson 1996: 9; Wright 1998: 10; Nugent and Raisinghani 2002: 8; Smith 2006);
- design and provision of software for countering 'hacking' or unauthorised intrusion (e.g. firewalls, intrusion-detection systems, early-warning systems) (Grow 2004: 84; Grabosky and Smith 2001: 40);
- design and provision of software for detecting and eradicating 'malicious software' (e.g. viruses, worms and Trojan horses);
- provision of systems for safeguarding confidential, proprietary and business-sensitive data (e.g. encryption software to enable secure financial transactions over public networks such as the internet, and technologies preventing unauthorised copying and reproduction of copyright-protected digital content such as software, music and motion pictures) (Rassool 2003: 5–6; Vaidhyanathan 2003: 176–177);
- training for organisations and their employees in using and implementing security systems and procedures.

The overall financial scale of this industry is difficult to determine with precision, depending on how narrowly or broadly 'computer crime control' is defined. Nevertheless, we can glean some preliminary indications of its extent and growth. Research estimates placed US companies' spending on computer security at $2.8 billion in 1999, $3.4 billion in 2000, projected to rise to $9.9 billion in 2005 (Rombel 2001). The global financial outlay on such products and services was placed at $27 billion (Grow 2004: 84). Overall growth in the sector has remained high (30 per cent in 2003 – *Computer Weekly* 2003: 1), bucking the slowdown in much of the IT sector following the bursting of the 'dot.com' bubble (Castells 2002: 105–106). The market for security software (such as firewalls, anti-virus systems and intrusion-detection systems) has expanded particularly strongly, growing by 18 per cent between 2001 and 2002 (Lemos 2002) and continuing on a strong upward trend. The market for managed security services (wherein organisations outsource computer security provision to an external contractor) has also shown strong growth in the early years of the new millennium, with an estimated expansion from $720 million in 2000 to $2.2 billion in 2005 (Lemos 2002).

The growth of the CCCI has paralleled the rapid expansion of networked computing (especially the internet). Between 1994 and 2008 the number of countries connected to the internet increased from 83 to more than 200 (Furnell 2002: 7; GWE 2008). In December 1995 there were an estimated 16 million internet users worldwide; by November 2007, this figure had risen to 1.26 billion, some 20 per cent of the world's total population (IWS 2007). The figure is predicted to rise to 2 billion by 2010 (Castells 2002: 3). During this period, there has been a rapid increase in the number of computer-enabled and computer-focused crimes, or cyber-crimes, and a corresponding increase in the costs incurred as a result. For example, in 1998 the FBI reported that computer intrusion incidents had increased 250 per cent over a two-year period (Lilley 2002: 32); in Russia, computer crime is estimated to be increasing at 400 per cent per annum (Saytarly 2004); hacking is estimated to have cost businesses $1.6 trillion in 2000 alone (Newman and Clarke 2003: 55). Such alarming figures have helped drive demand for greater computer security; indeed, the CSI has not been slow in exploiting the fear of cyber-attack to stimulate demand for its products and services. Taylor (1999: 214) cites Marx's claim that 'crime, through its constantly new methods of attack on property, constantly calls into being new methods of defence' and goes on to suggest that, in the contemporary context, crime must be viewed as a driver of innovation and expansion in the area of security provision (ibid.: 222). Such an analysis seems particularly apposite in relation to computer crime, where the threat of victimisation has sharpened the appetite for computer security amongst businesses, governments and individual computer users alike. In the next section I move to examine why this expansion has, to a significant extent, taken the form of commercially driven private provision, creating a burgeoning industry in computer crime control.

Neo-liberal governance and crime control

There now exists a voluminous literature dealing with recent changes in policing and crime control, and numerous attempts to situate such changes within a wider theorisation of social, political and economic transformation. I will not engage here with the full range of such changes (such as the rise of 'actuarial justice' and 'penal expansionism') as they have little direct relation to the issue at hand (on the former, see Feeley and Simon 1992, and on the latter Christie 2000). Nor can I evaluate the wide range of social-theoretical frameworks mobilised to explain such changes, such as the 'risk society' thesis, or the supposed transition to a 'late' or 'post' modernity (on risk see Loader and Sparks 2002: 92–95; on late modernity and crime control see Young 1999 and Garland, 2001; on post-modernity and policing see Reiner 1992). Indeed, there would appear to be considerable disagreement about the scope and scale of such changes and their wider significance. Thus, for example, Bayley and Shearing (1996) claim that there has occurred a fundamental fragmentation and pluralisation of policing, and that the current trends mark the end of the monopolistic system of public policing established in the early nineteenth century; Jones and Newburn (2002), in contrast, adopt a more gradualist approach, seeing such changes as extensions of a long-established process in which social control is 'formalised' by both public and private agencies. However, what most analysts seem to agree upon is that there *have* been significant shifts in the provision of policing and crime control, marked by its commodification in tandem with a 'responsibilisation' in which the burden of crime control is shifted towards non-state agencies and individuals (Muncie 2005: 37, 39). In this context, non-state actors are encouraged to protect themselves against the threat of criminal victimisation via market mechanisms, such that they become *consumers* of security good and services rather than *citizens* who have a right to expect protection from the state as part of a social contract (Bowling and Foster 2002: 981–982).

These trends, I suggest, typify the organisation of computer crime control and policing. Responsibility for crime prevention falls largely (albeit not exclusively) upon the potential victims, and they are typically expected to access such provision through the purchase and contracting of security good and services. Likewise, the provision of crime prevention and detection constitutes an ever-expanding array of market opportunities for private security providers. I wish to situate these developments in computer crime control within the context of systemic political-economic changes, specifically the emergence of new modes of social coordination, or governance, as part of the transition to a neo-liberal mode of capitalism. This is not to suggest that accounts which explain changes in crime control with reference to an 'epochal' transformation of social formation (e.g. the end of 'modernity') are without useful insights; however, I hold that the level of analytical generalisation at which such accounts operate makes it difficult to furnish adequate explanations of more specific, localised phenomena (for an elaboration of this critique, see Penna and Yar 2003; Yar and Penna, 2004).

The concept of governance has become one of the most oft cited yet contested social scientific concepts in recent years (Lee 2003: 3). On my understanding, governance refers to a specific mode of social coordination and ordering, one which moves away from 'government' by a centralised state apparatus and towards a more heterodox, self-organizing *network* of actors situated within market and civil societal, as well as governmental and quasi-governmental spheres (Rhodes 1997). In the move to governance the state increasingly eschews the hierarchical, top-down delivery of social order, instead externalising responsibility onto extra-governmental actors, while maintaining an interest in 'steering' such activity through the formulation of policy goals and agenda setting (Crawford 2006). I see the emergence of this mode of social governance as integrally tied to the transition to a neo-liberal regime of capitalist accumulation and regulation (Jessop 2002). The 'failures' of Keynesian economics and state welfarism to ensure either sustained economic growth or the achievement of public welfare goals (including crime reduction) paved the way for a 'de-centring' of the state. This move found its political articulation in the rise of the New Right, with its insistence that the competitive mechanisms of the market are inherently more efficient than state bureaucracies in the delivery of social goods; that the scope of state expenditure needs to be 'rolled back' in order to reduce the burdens of taxation which inhibit economic growth; that the state provision of social goods undermines individual self-reliance and responsibility, and curtails freedom of choice. Consequently, there has emerged a range of state strategies in consonance with the neo-liberal project, including liberalisation and deregulation of markets, and privatisation of the public sector (Jessop 2002: 461). In the domain of policing and crime control, the displacement of responsibility onto extra-governmental actors (businesses, voluntary organisations and individuals) has the advantage of externalising costs and relieving public agencies such as the police from the burden of responsibility for an ever-widening array of crime-control tasks.

I argue that the emergence and rapid growth of the commercial sector in computer crime control must, in the first instance, be situated within the context of the neo-liberal regime of governance. As Wall (2001: 174) notes, while the public police may resent the consolidation of the private sector in cyber-crime policing, 'resource managers appear happy not to expend scarce resources on costly investigations'. The character of this tension becomes apparent if we consider the limited resources the police can make available for tackling the rapid rise in computer-related crimes. For example, the UK police's National Hi-Tech Crime Unit was established in 2001, comprising 80 dedicated officers and with a budget of £25 million; however, this amounts to less than 0.1 per cent of the total number of police and less than 0.5 per cent of the overall expenditure on 'reduction of crime' (Home Office 2002; Wales 2001: 6). As Wall (2007: 183) observes 'the public police mandate prioritises some offending over others', and the public and political focus upon 'street crimes' and 'conventional crimes' (crimes of violence, robbery, drug-related offences and such like) diverts resources away from the policing of computer crime. In such a situation, the police have little choice but to accept the incursion of commercial organisations into computer crime control. The role of the commercial sector goes beyond the provision of crime-prevention services and technologies and increasingly takes the form of a range of investigatory and enforcement functions (in other words, it spans reactive as well as proactive or preventative interventions). For example, in the area of anti-piracy and the protection of digital rights, recent years have seen the proliferation of a range of private organisations funded by commercial clients seeking protection of their intellectual property from illegal exploitation. Examples of such organisations include the Counterfeiting Intelligence Bureau, the International Intellectual Property Alliance, the International Anti-Counterfeiting Coalition, the Alliance Against Counterfeiting and Piracy, the Coalition for Intellectual Property Rights, the Artists Coalition Against Piracy, the Anti Counterfeiting Group and the Federation Against Copyright Theft. Such organisations purport to 'lift the burden of investigation from law enforcement agencies' (AACP 2002: 2) by engaging in a range of increasingly intensive policing activities. Such policing services for the potential victims are available on the basis of ability and willingness to pay in the form of subscriptions and membership fees.

All of the foregoing is not meant to suggest, however, that the emerging structure of computer crime control ought be viewed as a straightforward displacement of responsibility from the public (state) to the private (market) domain. Rather, as already noted, implicated within the new mode of governance is a range of actors that can be situated across public, quasi-public and private spheres (what Brenner (2006) calls a system of 'distributed policing'). Thus, in common with developments in crime control more broadly, computer and information security is now provided by non-market quasi-public regulatory bodies (such as the Internet Watch Foundation) and communities of self-policing (such as the Association of Sites Advocating Child Protection – ASCAP), in addition to commercial provision. Consequently, the language of 'privatisation' cannot in and of itself adequately grasp the complex array of computer crime control mechanisms that spans hierarchies, markets and networks, and which increasingly blurs the boundaries of public and private. These arrangements can perhaps be better understood in terms of what Bayley and Shearing (2001) call 'multilateralisation'. Nevertheless, insofar as commercialised provision comprises a key (and rapidly expanding) element of this apparatus, the examination of its emergence as a market is both warranted and required.

The information economy and intellectual property

A further crucial driver in the growth of the computer crime control industry can be located in the transformation of economic activity brought about by the development of an informational economy. Recent academic discussions have identified a transformation of economic life, often associated with the emergence of 'post-industrial' capitalism (Bell 1999). Since the 1970s, western economies increasingly have moved away from their traditional dependence upon industrial production, and economic growth has come to depend upon the creation, exploitation and consumption of information. As Castells (2002) points out, economic dependence upon information (typically in digitised and computerised form) has come to permeate economic activity, being central to a wide range of business activities – research and development, product design, coordination of manufacturing, and advertising, sales and marketing. The array of financial services and transactions essential for the working of the capitalist economy has also shifted into the electronic information communication environment; prime examples include the computerised working of contemporary trading in stocks and shares, money and futures markets, banking and insurance. From the perspective of consumers, the development of information technology has reshaped the experience of shopping, with goods and services being accessed and purchased via electronic communication, such as 'e-shopping' and online banking via the internet. The inevitable upshot of these developments has been a great dependence upon those electronic systems that store, process and communicate information. The potential threats to these systems' integrity, be it the risk of unauthorised access, manipulation, corruption or destruction, have served to create a new market for computer and information security, as both business organisations and consumers seek to protect themselves from the vulnerabilities associated with their use of ICTs. As the UK government's Department of Trade and Industry (DTI) notes: 'Protecting information has never been more important. Organisations face a wide range of risks to their data, including virus attacks, inappropriate usage, unauthorised access and theft or systems failure' (DTI 2004: 1). It goes on to note how theft of information has profound implications in terms of commercial losses (business-sensitive and proprietorial information may fall into the hands of competitors), damage to reputation and trust, and financial costs associated with potential legal action and data recovery (DTI 2004: 9). Consequently, state strategies aimed at maximising economic gains from the information economy place great emphasis upon encouraging business organisations to secure computer systems and data storage against illegal intrusion and interference.

A second dimension of this 'informationalisation' of the capitalist economy has been the generation of profit through the production and consumption of 'pure' informational goods. In the information economy, the so-called 'culture industries' comprise key sectors. Such businesses produce goods that basically are made up of various distinctive arrangements of digitised information – prime examples include computer software, music and motion pictures. The capacity to generate profits from such 'intangible goods' requires that their producers and owners are able to exercise proprietary control over their circulation. Legally speaking, such rights and controls are enshrined through intellectual property laws. Such laws prohibit the unauthorised and unremunerated copying and distribution of original forms of expression (Coombe and Herman 2004: 561). However, the digitisation of such goods, in combination with the internet that permits information to be transmitted and shared on a worldwide scale, threatens producers' and owners' control. Thus, for example, in the area of musical recordings, the International Federation of Phonographic Industries (IFPI) claims that the global production of 'pirated' recordings now amounts to 1.8 billion units per annum, the bulk in the form of CDs; this means that one in three CDs sold is an unauthorised copy. In financial terms, this amounts to $4.6 billion (IFPI 2003: 2). These figures cover only commercial piracy and do not include copying by consumers and distribution via peer-to-peer 'file sharing' internet sites. It is claimed that 81.5 million people (4.98 per cent of the world's internet users) illegally downloaded music in the course of 2003. This piracy is deemed to have led to an average *monthly* loss of $450 million to copyright holders throughout 2004 (DIG 2004). In the area of computer software, global losses from piracy were pinned at $13.08 billion for 2002 (BSA 2003: 3). Similarly, in the area of motion pictures the Motion Picture Association of America (MPAA) claims that the US film industry loses in excess of $3 billion per annum worldwide as a result of piracy (MPAA 2005). In 2002, over 7 million pirate DVDs were seized worldwide (Valenti 2003). The MPAA further estimates that more than 350,000 unauthorised internet movie downloads take place every day (Valenti 2002). If true, this would amount to a staggering *125 million* film downloads per annum. This scale of financial losses has been deemed to threaten the long-term viability of key sectors in the knowledge economy. Consequently, ever more elaborate security measures have been deemed necessary to protect informational good against piracy, thereby generating extensive market opportunities for computer security companies which provide appropriate 'solutions' (such as anti-copying technologies and means of identifying legitimate products through the embedding of 'digital signatures' and the like). In short, the greater the economic expectations placed upon informational goods, the greater the sense of risk and insecurity, and the greater the demand for tools and services which promise to secure such property against theft.

9/11 AND THE SPECTRE OF CYBER-TERROR

A third and final driver behind the computer crime control industry can be located at the more conjunctural level of politics. In the course of the 1990s, western societies' use of information technologies generated in political, state-security and military arenas a heightened sense of vulnerability. The basic claim here is that western advanced nations' social and economic stability is now crucially reliant on what has been dubbed the 'critical information infrastructure' (CII), and that serious disruption to this infrastructure could result in potentially catastrophic consequences (Dunn and Wigert 2004). Thus, for example, accidental failure or deliberate sabotage of the computer systems governing financial markets could induce a massive economic crisis. Moreover, it has been noted that the complex material infrastructure of western nations has been incrementally integrated with computerised systems, such that the basic coordination and functioning of water, power, transport and emergency services is now reliant on electronic communications (Milone 2003). The failure of these information systems could thus induce effects ranging in severity from mild inconvenience to serious loss of civilian life. Consequently, security analysts identified a growing threat from 'information warfare' – warfare conducted by targeting information rather than material or human assets. At the same time, it was held that computer-related vulnerabilities might make such warfare an appealing option for 'terrorist' actors; as one commentator has put it, a scenario in which the 'logic bomb' displaces the 'truck bomb' as a weapon of choice (Denning 2000).

Such threat assessments have stimulated extensive investment in programmes to secure information infrastructures against cyber-attack, especially in the wake of the 9/11 terrorist attacks. In 1999, the Clinton administration committed US$1.46 billion for combating the threat of cyberterrorism (Hamblen 1999; Miyawaki 1999). In 2000, they issued the first comprehensive plan for protecting the US critical information infrastructure, *Defending America's Cyberspace: National Plan for Information Systems Protection*. In the wake of 9/11, the Bush administration earmarked a further US$839.3 million for cyber-security as part of the Homeland Security appropriations bill (Dunn and Wigert 2004: 201). Similarly, other nations have substantially increased the resources allocated to protecting information infrastructures in the wake of the 9/11 attacks: for example, Australia's CIIP budget was set at US$2 million in May 2001 but tripled the following year to US$6 million, and allocated a total of US$24.9 million over four years (Dunn and Wigert 2004: 43). The rapid expansion of CII protection budgets, fuelled by a heightened sensitivity to potential terrorist attacks, has boosted demand for computer security services, thereby driving the growth of the computer crime control industry over recent years. This upsurge in anti-terrorist computer security provision can be viewed as illustrative of a broader dynamic in which fears of crime risks (real or imagined) actively incite demand for innovation and expansion in crime control – an instance of what Hope (2006) calls 'reflexive securitisation', a kind of feedback loop in which social action and reaction are shaped by assessments of 'the crime problem'.

THE COMPUTER CRIME CONTROL INDUSTRY: PROPOSALS FOR AN INTEGRATED RESEARCH AGENDA

In a comprehensive review of existing criminological literature on the transformation of policing and crime control, Bayley and Shearing (2001) propose a systematic programme of focused enquiry into current developments. In concluding I suggest their agenda can be usefully adapted to chart a parallel path for examining the market-oriented provision of computer and information security, a phenomenon that, as already noted, has thus far been largely neglected in the otherwise voluminous criminological literature on the new landscape of crime control. Such an agenda ought to entail, first, a careful empirical description of the range of actors implicated in the delivery of computer crime control and the ways in which the markets for their goods and services are organised, developed and used by their 'consumers'. Second, it ought to examine the impact of such developments on public policing, and the ways in which public crime-control strategies both shape and respond to these changes. This would include an investigation of both the potentialities for effective cooperation and coordination amongst actors and a critical evaluation of the tensions and inefficiencies to which these arrangements might give rise (instances of what Jessop (1999) calls 'governance failure'). Third, it needs to address the social impacts of these developments in terms of the effectiveness of commercial provision, as well as the issues of equity and accountability that inevitably arise when dealing with provision of security that is socially dispersed and delivered according to the economic capacity of individualised market actors. Taken together, such enquiries would, it is hoped, furnish valuable insights into this important area of crime control and help inform a regulatory framework that balances economic opportunity with efficiency, equity, accountability, security and liberty.

CONCLUSION

In this chapter I have attempted to trace the factors that have fuelled the rapid growth of a private sector in computer crime prevention and control. This, I have argued, can be situated in the first instance within the trend towards the 'securitisation' of everyday life across myriad social domains. Discourses of insecurity incite ever more concerted attempts to restore a situation in which risks were supposedly less malevolent and prominent, and ordered regularities of action were more apparent. I have further suggested that the expansion of the computer crime control industry can be viewed as part of the broader reconfiguration of crime control and policing. Across many western industrial societies, the move to neo-liberal modes of governance has resulted in the emergence of non-state-centred networks that provide crime control in the form of market-mediated goods and services. In this regime, the responsibility for securing society against the threat of criminal predation falls upon 'responsibilised' individuals and organisations rather than the state and its formal crime-control agencies. However, beyond these more general developments, a number of specific social, economic and political changes have served to expand market opportunities for private provision in computer and information security. First, the greater economic dependence upon information communication technologies and informational goods has resulted in an imperative to secure such systems and resources against unauthorised exploitation and manipulation. Second, political sensitivities around the potential impacts of terrorist attacks have inspired efforts to secure critical information infrastructures against cyber-attacks. Taken together, these drivers have helped to establish the computer security industry as a major, and growing, element in the private provisions of crime prevention and control. This chapter is intended as a provisional mapping of the terrain of this industry and therefore concludes with proposals for a systematic and integrated agenda for further research.

References

AACP (Alliance Against Counterfeiting and Piracy) (2002) *Proving the Connection: Links between intellectual property theft and organised crime*, London: AACP.

Bayley, D. and Shearing, C. (2001) *The New Structure of Policing: Description, conceptualization and research agenda*, Washington: National Institute of Justice.

Bayley, D. and Shearing, C. (1996) 'The future of policing', *Law and Society Review*, 30(3): 585–606.

Bell, D. (1999 (orig. 1973)) *The Coming of Post-Industrial Society*, New York: Basic Books, 3rd edition.

Bowling, B. and Foster, J. (2002) 'Policing and the police', in M. Maguire, R. Morgan and R. Reiner (eds) *The Oxford Handbook of Criminology*, Oxford: Oxford University Press, 3rd edition.

Brenner, S. (2006) 'Cyber-crime: Rethinking crime control strategies', in Y. Jewkes (ed) *Crime Online*, Cullompton: Willan.

BSA (Business Software Alliance) (2003) 'Eighth Annual BSA Global Software Piracy Study', at: http://www.bsaa.com.au/downloads/BSA_Piracy_Booklet.pdf

Burgess, A. (2004) *Cellular Phones, Public Fears and a Culture of Precaution*, New York: Cambridge University Press.

Castells, M. (2002) *The Internet Galaxy: Reflections on the internet, business, and society*, Oxford: Oxford University Press.

Christie, N. (2000) *Crime Control as Industry*, London: Routledge.

Computer Weekly (2003) 'IDC Sees Bright Future for IT Security Services', available at http://www.computerweekly.com/Articles/2003/04/30/194225/Idc-sees-bright-future-for-it-security-services.htm (consulted March 2005).

Coombe, R. and Herman, A. (2004) 'Rhetorical virtues: Property, speech, and the commons on the world wide web', *Anthropological Quarterly*, 77(3): 559–573.

Crawford, A. (2006) 'Networked governance and the post-regulatory state? Steering, rowing and anchoring the provision of policing and security', *Theoretical Criminology*, 10(4): 449–479.

Denning, D. (2000) *Cyberterrorism*, Testimony before the Special Oversight Panel on Terrorism Committee on Armed Services U.S. House of Representatives, 23 May 2000.

Digital Intelligence Centre (DIG) (2004) *Digital Intelligence Centre: Archived News*, URL (consulted March 2005): http://www.itic.ca/DIC/News/archive.html#2004-06-09

DTI (Department of Trade and Industry) (2004) *Information Security: Hard facts*, London: DTI.

Dunn, M. and Wigert, I. (2004) *International CIIP Handbook: An inventory and analysis of protection policies in fourteen countries*, Zurich: Swiss Federal Institute of Technology.

Ericson, R.V. (2007) *Crime in an Insecure Word*, Cambridge: Polity.

Ericson, R.V., McMahon, M. and Evans, D. (1987) 'Punishing for profit: Reflections on the revival of privatisation in corrections', *Canadian Journal of Criminology*, 28(4): 355–387.

Feeley, S. (1991) 'The privatisation of prisons in historical perspective', *Criminal Justice Research Bulletin*, 6(2): 1–10.

Feeley, S. and Simon, J. (1992) 'The new penology: Notes on the emerging strategy of corrections and its implications', *Criminology*, 30(4): 449–474.

Furedi, F. (2007) *Invitation to Terror: The expanding empire of the unknown*, London: Continuum.

Furedi, F. (2006) *Culture of Fear Revisited*, London: Continuum.

Furedi, F. (2005) *Culture of Fear: Risk-taking and the morality of low expectation*, London: Continuum.

Furnell, S. (2002) *Cyber-crime: Vandalizing the information society*, London: Addison Wesley.

Garland, D. (2001) *The Culture of Control: Crime and social order in contemporary society*, Oxford: Clarendon.

Glassner, B. (2000) *The Culture of Fear: Why Americans are afraid of the wrong things*, New York: Basic Books.

Grabosky, P. and Smith, R. (2001) 'Telecommunication fraud in the digital age: The convergence of technologies', in D. Wall (ed) *Crime and the Internet*, London: Routledge.

Grow, B. (2004) 'Software', *Business Week*, 21: 84.

GWE (Global Web Explorer) (2008) 'How many countries are linked on the World Wide Web?' (consulted January 2008): http://www.guernsey.net/~sgibbs/www.html

Halverson, G. (1996) 'As internet booms, so do hacker-proofing measures', *Christian Science Monitor*, 88(144): 9.

Hamblen, M. (1999) 'Clinton commits $1.46b to fight cyberterrorism', available at http://www.cnn.com/TECH/computing/9901/26/clinton.idg

Home Office (2002) 'Home Office Annual Report 2001–2', available at http://www.homeoffice.gov.uk/docs

Hope, T. (2006) 'Mass consumption, mass predation: Private versus actors?', in

R. Lévy, L. Mucchielli and R. Zaubermann (eds) *Crime et Insécurité: Un demi-siècle de bouleversements – mélanges pour et avec Philippe Robert*, Paris: Hatmattan.

IFPI (International Federation of Phonographic Industries) (2003) *The Recording Industry Commercial Piracy Report 2003* (consulted March 2005): http://www.ifpi.org/site-content/library/piracy2003.pdf

IWS (Internet World Statistics) (2007) *Internet Usage Statistics: The internet big picture* (consulted January 2008): http://www.internetworldstats.com/stats.htm

Jessop, B. (2002) 'Liberalism, neoliberalism, and urban governance: A state-theoretical perspective', *Antipode*, 34(4): 452–472.

Jessop, B. (1999) 'The dynamics of partnership and governance failure', in G. Stoker (ed) *The New Politics of Local Governance in Britain*, Oxford: Oxford University Press.

Johnston, L. (1992) *The Rebirth of Private Policing*, London: Routledge.

Jones, T. and Newburn, T. (2001) 'The transformation of policing? Understanding current trends in policing systems', *The British Journal of Criminology*, 42(1): 129–146.

Kline, S. (2004) *Fast Food, Sluggish Kids: Moral panics and risky lifestyles*, Cultures of Consumption Working Paper No. 9, available at www.consume.bbk.ac.uk/working_papers/Kline_working_paper.doc

Lee, M. (2003) 'Conceptualizing the new governance: A new institution of social coordination', paper presented at the Institutional Analysis and Development Mini-Conference, 3 and 5 May 2003, Workshop in Political Theory and Policy Analysis, Indiana University, Bloomington, Indiana, USA.

Lemos, R. (2002) 'Computer-security industry leads the way to growth', *CNET*, 2 February, available at http://www.zdnet.co.uk/business/0,39020645,2103776,00.htm

Lilley, P. (2002) *Hacked, Attacked and Abused: Digital crime exposed*, London: Kogan Page.

Lilly, J. and Knepper, P. (1992) 'The corrections-commercial complex', *The Howard Journal*, 31(3): 174–191.

Loader, I. (2006) 'Fall of the "platonic guardians": liberalism, criminology and political responses to crime in England and Wales', *British Journal of Criminology*, 46(4): 561–586.

Loader, I. and Sparks, R. (2002) 'Contemporary landscapes of crime, order and control: Governance, risk and globalisation', in M. Maguire, R. Morgan and R. Reiner (eds) *The Oxford Handbook of Criminology*, Oxford: Oxford University Press, 3rd edition.

Manning, P. (2007) 'The symbolic framing of drug use in the news: Ecstasy and volatile substance abuse in newspapers', in P. Manning (ed) *Drugs and Popular Culture: Drugs, media and identity in contemporary culture*, Cullompton: Willan.

Milone, M. (2003) 'Hacktivism: Securing the national infrastructure', *Knowledge, Technology & Policy*, 16(1): 75–103.

Miyawaki, R. (1999) *The Fight Against Cyberterrorism: A Japanese view*, Washington: Centre for Strategic & International Studies.

Moore, D. and Valverde, M. (2000) 'Maidens at risk: "Date rape drugs" and the formation of hybrid risk knowledges', *Economy & Society*, 29(4): 514–531.

MPAA (Motion Picture Association of America) (2005) 'Anti-piracy', available at http://www.mpaa.org/anti-piracy/ (consulted March 2005).

Muncie, J. (2005) 'The globalisation of crime control – the case of youth and juvenile justice: Neo-liberalism, policy convergence and international conventions', *Theoretical Criminology*, 9(1): 35–64.

Newman, G. and Clarke, R. (2003) *Superhighway Robbery: Preventing e-commerce crime*, Cullompton: Willan Press.

Nugent, J. and Raisinghani, M. (2002) 'The information technology and telecommunications security imperative: Important issues and drivers', *Journal of Electronic Commerce Research*, 3(1): 1–14.

Penna, S. and Yar, M. (2003) 'From modern to post-modern penality?', *Theoretical Criminology*, 7(4): 469–482.

Rassool, R.P. (2003) 'Antipiracy – trends and technology (a report from the front)', available at http://www.broadcastpapers.com/asset/ IBCWidevineAntipiracy.pdf

Reiner, R. (1992) 'Policing and postmodern society', *Modern Law Review*, 55/56: 761–781.

Rhodes, R. (1997) *Understanding Governance: Policy networks, governance, reflexity and accountability*, Bristol: Open University Press.

Rombel, A. (2001) 'Internet security in an insecure world', *Global Finance*, (15)3: 28–32.

Saytarly, T. (2004) *Russia: Computer crime statistics*, 13 March 2004, available at http://www.crime-research.org/news/13.03.2004/131

Shearing, C. and Stenning, P. (eds) (1987) *Private Policing*, Newbury Park: SAGE.

Smith, R. (2006) 'Biometric solutions to identity-related cyber-crime', In Y. Jewkes (ed) *Crime Online*, Cullompton: Willan.

South, N. (1994) 'Privatizing policing in the European market: Some issues for theory, policy, and research', *European Sociological Review*, 10(3): 219–233.

Stroud, L. (2005) 'MMR – public policy in crisis: Whose tragedy?', *Journal of Health Organisation and Management*, 19(3): 252–260.

Taylor, I. (1999) *Crime in Context: A critical criminology of market societies*, Cambridge: Polity.

Vaidhyanathan, S. (2003) *Copyrights and Copywrongs: The rise of intellectual property and how it threatens creativity*, New York: NYU Press.

Valenti, J. (2003) *International Copyright Piracy: Links to organised crime and terrorism*, Testimony to The Subcommittee on Courts, The Internet, and Intellectual Property, US House of Representatives, 13 March 2003, URL (consulted March 2005): http://www.mpaaa.org/jack/2003/2003_03_13B.htm

Valenti, J. (2002) *If You Cannot Protect What You Own, You Don't Own Anything*, report presented to the US Senate Committee on Commerce, Science and Transportation, on behalf of the MPAA, 28 February 2002. Available at http://www.mpaa.org/jack/2002/2002_02_28b.htm

Wales, E. (2001) 'Global focus on cyber-crime', *Computer Fraud & Security*, 1, 6.

Wall, D. (2007) *Cyber-crime: The transformation of crime in the information age*, Cambridge: Polity.

Wall, D. (2001) 'Maintaining order and law on the internet', in D. Wall (ed) *Crime and the Internet*, London: Routledge.

Welch, M. (2006) *Scapegoats of September 11th: Hate crimes and state crimes in the war on terror*, New Jersey: Rutgers University Press.

Wright, H. (1998) 'Biometrics, the next phase in network security', *NZ Infotech Weekly*, 28 September, 10.

Yar, M. and Penna, S. (2004) 'Between positivism and post-modernity? Critical reflections on Jock Young's *The Exclusive Society*', *The British Journal of Criminology*, 44(4): 533–549.

Young, J. (1999) *The Exclusive Society*, London: SAGE.

Zedner, L. (2007) 'Pre-crime and post-criminology?', *Theoretical Criminology*, 11(2): 261–281.

Part V

(In)secure rights

Chapter 10

Technologies of surveillance and the erosion of institutional trust

Benjamin Goold[1]

INTRODUCTION

What are the costs of surveillance? What are the implications of allowing governments and private organisations to use sophisticated surveillance technologies such as CCTV, electronic tagging and data-matching software in their everyday dealings with individuals? At what point does routine, everyday surveillance become a threat to individual well-being, democratic values and the legitimacy of the modern state?

These questions are at the heart of many contemporary discussions about the meaning and significance of surveillance, and have in large part helped to animate the emerging field of surveillance studies. In attempting to answer these questions, writers have focused on a variety of potential dangers associated with the spread of surveillance, such as its impact upon individual privacy (Slobogin 2002; Goold, 2006) and its use as a tool for social and economic exclusion (Lyon 2007). Common to many of these accounts is concern with the individual costs of surveillance and with the threat certain technologies pose to individual civil liberties. Surveillance, it is argued, represents a threat to the individual because it threatens his privacy, identity and personal liberty.

It is true that the dangers posed to the individual by the expansion of surveillance are considerable. In this chapter, however, I take a slightly different approach to these questions and concentrate instead on the implications of surveillance for the organisations of government. In particular, I will examine the relationship between surveillance and what is referred to as 'institutional trust', and suggest that the increasing trend towards greater levels of state surveillance has the potential to undermine well-established norms of governance based on consent and a shared commitment to democratic forms of government.

1 I am indebted to the editors as well as Imogen Goold, Ian Loader and Lucia Zedner for their insightful comments on earlier versions of this chapter.

INSTITUTIONAL TRUST AND GOVERNANCE

Looking at the political studies literature, it is possible to identify two different and competing accounts of the relationship between trust and governance: cultural theories and institutional theories. While both theories regard trust as essential to the establishment and maintenance of democratic forms of government, their accounts of why trust matters and how it operates in practice are often radically different. Cultural theories regard trust as somehow socially embedded and transmitted from one generation to the next for reasons to do with a community's entrenched commitment to democratic values. By contrast, institutional theories treat trust as a product of institutional performance, capable of being enhanced or eroded according to the behaviour of government and government organisations.

Of the two approaches, it is the cultural theories that perhaps ascribe the greatest importance to matters of trust.[2] According to these theories, trust is crucial both in the sense of providing active political support for the democratic project (Almond and Verba 1963; Putnam 1993; Mishler and Rose 2005) and also in the more mundane sense of making everyday governance possible. Because individuals regard governmental institutions as democratically legitimate, they recognise their authority to exercise governmental power and (for the most part) willingly submit to that power. In this regard, cultural theories of democracy regard institutional trust as a form of social capital, a minimum amount of which is necessary for the proper and effective functioning of government.

As noted by Mishler and Rose (2005) in a recent discussion of the political consequences of trust, these cultural theories offer a variety of explanations of the specific relationship between trust and governance, and why societies have a vested cultural interest in producing it. First and foremost, trust is important because it serves to provide a minimum level of support for existing democratic structures. Put another way, trust helps to create and sustain a relationship between the government and the citizenry that makes it possible for the executive to pursue unpopular goals in the short or medium term without having to fear the withdrawal of democratic support or loss of legitimacy. Individuals accept decisions they may not like or agree with because of their general trust in government and belief that the institutions of government are generally committed to their welfare and well-being (Bianco 1994; Weatherford 1984). Second, trust matters because it helps to ensure that non-democratic alternatives to existing government structures are actively resisted by the public (Rose, Mishler and Haerpfer 1998). Trust in this sense acts as a bulwark against attempts to establish more authoritarian

2 This section draws extensively on the work of Mishler and Rose on institutional trust, and in particular on their excellent and nuanced summary of the differences between the culturalist and institutionalist positions (Mishler and Rose 2005).

modes of governance and helps to bolster the legitimacy of democratic government when it is under threat. Finally, it has been suggested that trust is important because it promotes political involvement and democratic participation (Norris 1999; Putnam 1993). As Mishler and Rose observe, trust in this way 'strengthens citizens' beliefs that government is responsive and encourages citizens to express their demands via participation in activities from voting to joining organisations' (Mishler and Rose 2005: 13).

While institutional theorists may acknowledge that these are all ways in which trust can operate in society, they differ from cultural theorists by suggesting that these values are not necessarily culturally determined, nor are particular forms of trust themselves cultural characteristics. Instead, they argue that trust is generated as a specific response to institutional performance, and that it is not so much transmitted across generations as continually reproduced in response to the actions of government (Jackman and Miller 1996; Mishler and Rose 2005). Seen in this way, trust is not so much an act of faith – necessary in order for the democratic project to work – but rather a rational reaction to the performance of government.

If one assumes for a moment that the culturalists are correct in their understanding of trust and its role in governance, then it follows that the individual actions of a government or government institution are – unless they are highly undemocratic – unlikely to shake long-term public trust. This is because the commitment to democracy and general trust in government runs deep and has a cultural foundation that is based on a long-term view of the relationship between government and the governed. In contrast, if the institutionalists are correct, then trust is a form of social capital that can be gained and lost relatively easily. The performance of individual governments and particular institutions can, according to this model, have a significant impact on the extent to which citizens are prepared to place their trust in the general framework of democratic government. In theory their failure can lead to the complete withdrawal of trust and the search for new systems of governance.

Of course, the reality of the relationship between governance and trust is most likely to lie somewhere between these two extremes. Although the actions of particular governments and institutions are likely to have an immediate effect on public feelings of trust, it is also likely that those actions will be viewed through a culturally determined lens that attempts to put breaches of trust in their proper historical, political and social context. Having said this, while it is reasonable to conclude that our reactions to various governance strategies that affect trust are likely to reflect a mixture of the culturalist and institutionalist views, what is clear is that trust matters to governance and, perhaps more importantly, it can be lost.

State surveillance and its implications for trust

If it is true that institutional trust is to some extent necessary for the exercise of government power in modern democratic societies, how is the growing use of surveillance technologies like public-area CCTV likely to affect the operation and reproduction of that trust? Unlike other forms of governance, surveillance is particularly interesting in the current context because its very existence implies some basic absence or withdrawal of trust – that is, it is a technique that is frequently used in situations where the state either does not trust the general public to behave well or feels that the mere threat of sanctions may not be enough to deter unwanted forms of behaviour. In this sense, the growing use of public surveillance technologies by governments in countries like the United Kingdom and the United States may signify the state's loss of trust in the public, and a repudiation of some aspects of the trust relationship as identified by both the cultural and institutional accounts discussed above.

Certainly, the shift from the individualised surveillance that characterised the first half of the twentieth century to the sort of mass surveillance identified by Marx, Lyon and others would seem to be indicative of this (Lyon 2001; Marx 1985). Whereas in the past the surveillance powers of the state were directed only at particular individuals who were deemed to be a risk or undeserving of trust, today it can be argued that those surveillance powers are instead directed at everyone. Public-area CCTV cameras are an especially good example of this. Although it can be argued that their primary function is to detect and deter specific instances of criminal conduct, in practice they have the effect of turning large swathes of public space into sites in which everyone is – at least potentially – regarded as suspicious and subject to mistrust. Equally, calls for a national DNA database and the adoption of a universal system of identification cards in the United Kingdom may in effect send a message to the public at large that everyone is open to suspicion. Government, it implies, has little need for trust when it has the tools of mass surveillance at its disposal.

What is perhaps paradoxical (or just simply ironic) about this expansion in the use of surveillance technologies by the state is that although it can be read as a withdrawal of trust in the public, the need for new technologies and techniques is usually justified by an appeal to trust in government. For example, concerns that a national identity card might be used in the future as the basis for some massive, intrusive database are dismissed by government ministers and officials on the ground that citizens should trust that such systems and information will not be misused or abused by either the current or future governments. Similarly, suggestions that CCTV cameras threaten individual privacy and autonomy in public spaces are met with the response that only those who have something to hide have something to fear, despite the fact that at least initially all are treated as subjects of suspicion. In this sense, the recent expansion in the use of surveillance technologies in many democratic states has been marked by what might be called an asymmetry of trust – that is, the untrusting state asks the public to trust it as it expands the apparatus of suspicion and surveillance.

It is against this general background that the rest of this chapter considers the potential implications of this asymmetry of trust for the future of governance in states such as the UK and the USA. If surveillance represents a particular challenge to the foundations of institutional trust, what might the implications of a continued expansion in the surveillance apparatus of the state be? What are the likely costs in terms of the capacity of the state to govern effectively, and what might be done to prevent a radical erosion of trust on the part of the public and weakening of state legitimacy?

The withdrawal of public trust?

It has been noted above that the increased use of surveillance technologies might send a particularly negative message to members of the public about how the state views them and the extent to which they can expect the state to trust them. Attempting to predict how individuals and the public might react to this withdrawal of trust (and the possible implications for governance) requires us to look more carefully at the conditions under which surveillance power is acquired and exercised by the state. Taking the divide between the culturalist and institutionalist positions as a starting point, it may be useful to deal separately with the general and specific implications of surveillance for public trust.

From a culturalist perspective, it is clear that states' growing use of surveillance technologies against their populaces has the potential to undermine the generally held trust in government within a society. As a number of writers have noted, the ability of a society to resist moving towards more authoritarian forms of government relies in part on a general belief that democracy – and the implicit value it gives to institutional trust – is a good thing (Gibson, Calderia and Baird 1998; Gibson and Calderia 2003). By engaging in widespread forms of public surveillance, it can be argued that the state risks undermining the normative commitment of citizens to democratic government and with it their commitment to associated values such as the protection of individual rights and civic responsibility. More simply put, we can speculate that one response to the withdrawal of trust in the public by the state might be a withdrawal of trust in the state by the public. To some extent, this is a point that has already been made by a number of criminologists writing on risk, uncertainty and the tactic of responsibilisation in late modern states (Garland 2002; Young 1999). As the state has sought to make individuals more responsible for their own protection and to develop a greater understanding of the risks they face, the state has in turn undermined its claims to being the sole provider of order and (in extreme cases) to its monopoly on the use of force. The rise of private policing and gated communities in countries like the United States in part represents a response by the public to the state's withdrawal from certain areas of law enforcement (Low 2004; Simon 2007). In a similar way, it is possible to imagine a situation where the public may also withdraw their trust in certain institutions of government and with it their willingness to be governed at all. If we assume that government authority is exercised by consent and 'on trust' – that is, the public abide by the decisions of government agencies not because of any threat of sanction but because of a general belief that the institution is acting in their interests (Tyler 2006) – then it follows that the withdrawal of trust may have serious implications for how government agencies go about their business.

Perhaps the most obvious example of an agency that would suffer greatly from a loss of trust is the police. In the United Kingdom in particular, the notion of 'policing by consent' is one that has underpinned the exercise of police power for over a century. The police do not, for example, carry firearms or routinely resort to violence in order to assert their authority because they assume – for the most part correctly – that they operate with the support and trust of the general public. If, however, that trust was gradually withdrawn in response to oppressive police surveillance, then it is easy to see how the everyday work of the police would quickly become considerably more difficult.[3] Aside from the fact that the police may find that their authority is no longer taken for granted, they may also find that their major source of information, the public, are no longer willing to be as cooperative or forthcoming. If this scenario seems somewhat unlikely, one need only look at the relations between the police and certain minority communities in the United States, where the relationship of trust is all but extinct. For police officers working in the poor, urban areas of cities like Detroit, Washington DC and Baltimore, trust is an extremely rare commodity, and as a result their work is often dangerous and hampered by a lack of public support.

In a similar way, one might also expect recent attempts in the United Kingdom to strengthen airport security through the increased use of surveillance technologies to have an impact on relations of trust between the public and agencies such as the Home Office and the Border and Immigration Agency (formerly known as the Immigration and Nationality Directorate). In applying broad, new security measures equally to all travellers, these institutions risk promoting a general atmosphere of mistrust in which individuals begin to see themselves as undifferentiated subjects of security rather than the beneficiaries of it. While in practical terms this may manifest itself as little more than a sense of dissatisfaction amongst the majority of travellers and members of the public visiting airports, the gradual erosion of trust in this environment may lead to more serious consequences. It is possible, for

3 This is an argument made some 20 years ago by Kinsey, Lea and Young (1986).

example, that increased levels of surveillance may cause individuals to focus more on ensuring that their otherwise benign behaviour does not attract unwanted and unfounded suspicion, while at the same time ignoring the behaviour of their fellow travellers. In this way, the withdrawal of trust and oppositional relationship created by the increase in surveillance may help to create a situation in which the overall amount of information being made available to those agencies concerned with airport security actually declines. Without a relationship of trust between the individual and the government institutions, the work of those concerned with surveillance and overall security may become that much harder.

Yet while it is reasonable to assume that the continued expansion in the apparatus of state surveillance may have negative effects on public confidence in government in the medium to long term, it is specific breaches in public trust arising from abuses of surveillance power that are likely to do the most damage to public trust. As has already been noted, many modern surveillance initiatives have been presented to the public accompanied by the message that government can be trusted not to abuse the technology in question or misuse the information it gathers. In such circumstances, it can be argued that the government creates a specific covenant of trust with the public, in addition to any general relationship of trust that might have existed already. A case in point is the United Kingdom's plan to introduce a compulsory national identification card. Speaking at the Institute of Public Policy Research in November 2004, the then Home Secretary David Blunkett dismissed concerns about the use of identity cards, suggesting that critics of the proposed scheme had fallen prey to an unfounded mistrust in the state based on the writings of a famed German philosopher:

> It was writers like Kant who first took the view that there is something suspicious about government activity, and that if a government is up to something, it must be about removing freedoms.

Interestingly, in the same speech Blunkett also appealed for greater public trust in the government on the grounds that, unlike the private sector, the government could guarantee that any data collected would be stored securely and be free from abuse. While at the time this appeal to trust was greeted with a degree of scepticism by civil liberties groups and large sections of the general public, in the light of recent events in the United Kingdom such claims seem even less believable. Almost four years after Blunkett's speech, it was revealed that the same government had lost two computer disks containing the personal details (including the names, addresses, dates of birth and bank account information) of 25 million people who had applied for state-funded child benefits in the previous three years. Even worse, in the months that followed the government was to suffer a number of other similar data losses, by the National Health Service (NHS) and the Ministry of Defence.

Taken together, the government's demand for greater surveillance powers and its very public failure to safeguard the sensitive information it already holds have the potential to seriously undermine public confidence and trust. Indeed, one might reasonably predict that one reaction to the recent round of data losses might be a resistance on the part of the public to providing sensitive information to the government in the future (Sunstein 2005). According to a poll conducted by MORI in November 2007 (following the loss of the child benefit data), 62 per cent of those surveyed felt that personal data held by the government was at risk.[4] While it is unlikely that such feelings will translate into a general mistrust of all aspects of government, it is reasonable to expect that they may make it harder for certain institutions to operate. Certainly, it would come as no surprise if in the years immediately following these data losses it was to emerge that fewer and fewer members of the public were willing to disclose sensitive, personal information to HM Revenue and Customs or their local hospital.

Looked at in these ways, it is apparent that the continued expansion of surveillance may have costs that go beyond questions of individual privacy and personal liberty.[5] Instead, surveillance has the potential to change the way in which citizens view and interact with public institutions, and the way in which they view and understand the operation of government and the exercise of governmental power. While it would be wrong to suggest – on the basis of the available evidence and recent trends at least – that the growing use of surveillance technologies by the state is likely to result in the complete withdrawal of public trust, even small cracks in the trust relationship are likely to have a significant effect on the operation of government institutions. This is especially the case in countries like the United Kingdom where levels of institutional trust and confidence in government have been high and stable over long periods of time. Although such public trust constitutes a valuable form of social capital, because it is so longstanding there is a danger that many of the institutions in which that trust resides have come to take it for granted and base a great deal of their authority on it.

Here again the British police are a case in point. Having never faced a serious challenge to their legitimacy, policing practices and policing culture have developed in an environment of general public trust over many generations. As a consequence, it can be argued that as an institution the police are particularly ill equipped to deal with situations in which they do not enjoy

4 For a discussion of the survey and its findings, see the discussion on ZDNET UK: http://www.news.zdnet.co.uk/security/0,1000000189,39291486,00.htm.

5 It is worth noting here that trust and privacy are also closely related. Although it is beyond the scope of this chapter to explore that relationship in any detail, insofar as privacy is often most strenuously asserted in situations in which trust has broken down, we might reasonably expect a rise in surveillance – and in state–public mistrust – to lead to calls for a strengthening of privacy rights and data-protection legislation. For a discussion of privacy and trust, see in particular works by Nock (1993) and Etzioni (2000).

that trust, as was briefly demonstrated by the events surrounding the 1984 miners' strike. To some extent, this is something that has been acknowledged by the police themselves. Although it is sometimes claimed that the police were behind efforts to expand the UK's public surveillance infrastructure in the 1990s, there is evidence to suggest that members of the Association of Chief Police Officers (ACPO) and many other senior officers were in fact concerned about the impact that CCTV might have on the relationship between the police and the public at large (Goold 2004). Trust, they recognised, was hard won and easily lost, and while the police were keen to reap the promised benefits of this new technology, they also wished to avoid becoming the focus of public suspicion.

Expanding surveillance and preserving trust

This chapter provides only an outline sketch of the possible relationship between surveillance and institutional trust. Clearly, in order to properly understand the impact that surveillance technologies and techniques may have on governance in modern democracies it is necessary to go beyond broad sociological observations and aim to acquire an empirically based understanding of how the public reacts to new surveillance measures and how this affects their view of government.[6] Having said this, in drawing attention to the potential institutional costs of surveillance, this chapter begs the question of whether anything can ever be done to avoid such costs. Is a loss of institutional trust an inevitable by-product of mass surveillance? Is there an inescapable trade-off between the level of state surveillance and the amount of trust the state can expect from the public? If not, then what can be done to ensure that trust is preserved in the face of some new expansion in the use of surveillance technologies?

In attempting to provide a framework for answering this question, it is useful to return to the opposing theories of institutional trust that were discussed at the beginning of this chapter. If we take a strict culturalist view, then it would seem that there is no obvious answer to any of these questions. Whether an increase in the state's use of surveillance technologies will lead to an erosion of public trust will depend on a host of cultural factors. It may be the case, for example, that other changes in the landscape of governance will

6 There are various ways in which researchers could empirically explore the relationship between surveillance and trust. A simple (albeit crude) approach would be to ask members of the public about their feelings of trust following the introduction of new, visible surveillance measures. A more productive approach, however, might be to look at public trust over time, both as a separate phenomenon and in conjunction with studies of the mundane interactions between individuals and particular government institutions. For example, one could check whether the recent data-loss scandals in the UK have affected how people interact with government agencies via the internet, or led to a fall in the number of tax returns or benefit claims filed online.

serve to counterbalance any loss of trust generated by the use of technologies like CCTV. Furthermore, we may also find that as members of the public become culturally conditioned to accept mass surveillance as a reality of modern governance, trust ceases to decline (even though the overall level of surveillance may continue to increase). In light of this, it is difficult to imagine a set of policies or practice that, if adopted, we could be sure would help to preserve and promote trust in the context of increasing state surveillance. Instead, what is demanded is a more general approach to the creation of trust, which focuses not on surveillance per se but rather the overall cultural conditions of governance in which it takes place.

If we are, however, inclined towards a more institutionalist understanding of the relationship between governance and trust, then the picture becomes somewhat clearer and perhaps more encouraging. As has already been noted, according to institutional theory trust is simply a rational response to institutional performance (Mishler and Rose 2005: 14). Whether surveillance has a negative impact on trust will therefore depend primarily on how it is carried out, and in particular whether the public regards its use as promoting or undermining those values they feel most committed to. Looked at another way, if the institutionalists are to be believed then it is not the establishment of DNA databases or the spread of CCTV that erodes trust but rather the inability of those institutions responsible for such surveillance to ensure that such technologies do not have an overly negative or unexpected impact on the political, social or economic life of the general public. What is required for the preservation of trust is not necessarily less surveillance but instead more and better regulations and safeguards.

CONCLUSION

If we assume that both the cultural and institutional theories of trust have some explanatory power, how should we then respond to the growing use of surveillance technologies by the state? Although much of the discussion in this chapter is speculative, a number of things are clear. First and foremost, there is a need for both individuals and governments to take a broader, more institutional view of the potential costs of surveillance. While academic commentators, civil libertarians and regulators have been right to focus much of their attention on the threat posed by technologies such as CCTV to privacy and individual liberty, surveillance is now so widespread that it may have larger consequences for institutional trust and the relationship between the citizenry and the state. If we fail to pay attention to the role played by trust in democratic society, then we risk ignoring or at the very least underestimating the true cost of surveillance. Second, it should be evident from the above discussion that we also need to develop a better understanding of the cultural foundations of trust, and in particular how these cultural forces condition our political and social reactions to new surveillance technologies. Finally, the above analysis suggests that it may also be profitable to take a broader view of the purpose of surveillance regulation. Whereas in the past regulators have tended to focus on protecting individuals from unnecessary or overly intrusive surveillance, if we are also concerned to promote institutional trust then a broader approach may be required. In particular, regulation may prove to be an effective means of changing existing governmental and institutional cultures, and ensuring that trust is seen as a potential cost of any expansion in surveillance or change in surveillance practice.

Of course, in practice it may be difficult to convince the public and policy makers that institutional trust is something that needs to be taken into account when deciding whether and in what ways we should allow the state to expand that apparatus of surveillance. Given that more familiar concepts such as privacy and identity have not always provided a sufficiently defined focus for discussions about the costs of surveillance, there is always the danger that drawing attention to the threat to trust may simply muddy the waters. Yet if our aim is to better understand (and predict) the impact of surveillance technologies on political life and our relationship to the state, we must at the very least begin to consider some of the questions raised in this chapter. If we fail to do so, then we risk exchanging institutional trust and a shared belief in the benevolence of democratic governance in favour of something altogether less optimistic and possibly more authoritarian.

References

Almond, G.A. and Verba, S. (1963) *The Civic Culture*, Princeton: Princeton University Press.

Bianco, W.T. (1994) *Trust, Representatives and Constitutents*, Ann Arbor: University of Michigan Press.

Etzioni, A. (2000) *The Limits of Privacy*, New York: Basic Books.

Garland, D. (2002) *Culture of Control: Crime and social order in contemporary society*, Oxford: Oxford University Press.

Gibson, J.L. and Calderia, G.A. (2003) 'Measuring attitudes toward the United States Supreme Court', *American Journal of Political Science*, 47(2): 354–367.

Gibson, J.L., Calderia, G.A. and Baird, V.A. (1998) 'On the legitimacy of national high courts', *American Political Science Review*, 92: 343–358.

Goold, B. (2006) 'Open to all? Regulating open street CCTV and the case for "symmetrical surveillance" ', *Criminal Justice Ethics*, 25(1): 3–17.

Goold, B. (2004) *CCTV and Policing: Public area surveillance and police practices in Britain*, Oxford: Oxford University Press.

Jackman, R.W. and Miller, R.A. (1996) 'The poverty of political culture', *American Journal of Political Science*, 40(3): 697–716.

Kinsey, R., Lea, J. and Young, J. (1986) *Losing the Fight Against Crime*, Oxford: Blackwell Publishing.

Low, S. (2004) *Behind the Gates: Life, security, and the pursuit of happiness in Fortress America*, New York: Routledge.

Lyon, D. (2007) *Surveillance Studies: An overview*, New York: Polity.

Lyon, D. (2001) *Surveillance Society: Monitoring everyday life*, Buckingham: Open University Press.

Marx, G.T. (1985) 'The surveillance society: The threat of 1984-style techniques', *The Futurist* (June), 21.

Mishler, W. and Rose, R. (2005) 'What are the political consequences of trust? A test of cultural and institutional theories in Russia', *Comparative Political Studies*, 38(1): 9–38.

Mishler, W. and Rose, R. (2001) 'What are the origins of political trust? Testing institutional and cultural theories in post-communist society', *Comparative Political Studies*, 34(1): 30–62.

Nock, S. (1993) *The Costs of Privacy: Surveillance and reputation in America*, Piscataway, NJ: Aldine Transaction.

Norris, P. (ed) (1999) *Critical Citizens: Global support for democratic governance*, Oxford: Oxford University Press.

Putnam, R.D. (1993) *Making Democracy Work*, Princeton: Princeton University Press.

Rose, R., Mishler, W. and Haerpfer, C. (1998) *Democracy and its Alternatives*, Baltimore: John Hopkins University Press.

Simon, J. (2007) *Governing Through Crime: How the war on crime transformed American democracy and created a culture of fear*, Oxford: Oxford University Press.

Slobogin, C. (2002) 'Public privacy: Camera surveillance of public places and the right to anonymity', *Mississippi Law Journal*, 72: 213–270.

Sunstein, C. (2005) *Laws of Fear: Beyond the precautionary principle*, Cambridge: Cambridge University Press.

Tyler, T. (2006) *Why People Obey Law*, Princeton: Princeton University Press.

Weatherford, M.S. (1984) 'Economic stagflation and public support for the political system', *British Journal of Political Science*, 14: 187–205.

Young, J. (1999) *The exclusive society: social exclusion, crime and difference in late modernity*, London: SAGE Publishing.

Chapter 11

Another side of the story

Defence lawyers' views on DNA evidence

Johanne Yttri Dahl[*]

Norwegian Minister of Justice, Knut Storberget, recently predicted a DNA revolution in Norway (Dagsavisen 24.07.2007). In December 2007, the Norwegian government decided to expand its forensic DNA database and granted 64 million kroner (approximately £5 million) to finance the 'DNA revolution'. According to Storberget (2007), DNA analysis is one of the most important tools available in the battle against criminality worldwide. In a press release, the Ministry of Justice (2007) stated that no method can outperform DNA, neither when it comes to efficiency nor credibility, and that it is necessary for the Norwegian police to have efficient tools like police elsewhere. Repeatedly, DNA advocates predict that DNA will contribute to increased detection of a variety of crimes, from volume crime, serious crime, organised crime, to national as well as international crime. Consequently increased use of DNA will free-up police resources. Moreover, the ability to detect more crime will contribute to increased levels of security (Storberget 2007). Similarly, the British Home Office (2004: 121) claims that its National DNA database (NDNAD) is revolutionising crime detection and that 'each week, the DNA Database identifies six murder suspects and matches over 700 profiles from crime scenes to named individuals' (ibid.).

However, not only politicians praise the advantages of forensic DNA evidence and DNA databases; police, media and the general public do so as well. DNA is often presented as a great new weapon in the police armoury in the war against crime, a tool that will lead to increased public security. DNA evidence is also talked of as a 'silver bullet' and the 'gold standard' for forensic identification. According to McCartney (2006: xii), '[p]ortrayals of the ineffability of DNA and its unrivalled ability to "solve" crime have led to determined effort and financial investment to significantly increase the use of DNA evidence in court and as a tool of detection through the NDNAD'. Because DNA is seen as reliable, trustworthy and secure evidence that will contribute to convicting the guilty and exonerating the innocent, DNA is

* I am grateful to Heidi Mork Lomell, Helene Oppen Gundhus, Katja Franko Aas, Ann Rudinow Saetnan and Erik Dahl for valuable comments on this chapter.

expected to increase security on a micro and a macro level; the individual's legal protection and the general rule of law. In short, a measure that will provide security in an insecure society. As Gerlach (2004: 33) states: '. . . DNA testing and banking are coming to be defined as secure biotechnologies that will produce enhanced security from dangerous criminals.'

This chapter, however, suggests that despite DNA being presented and perceived as a type of forensic evidence that will lead to increased security and justice, there exist important insecurities related to the use of DNA. There is an increasing body of literature on different aspects of insecurities regarding DNA evidence (see Diesen and Björkman 2003; Holmgren 2003; Koehler 2001; Lazer 2004; Lynch and McNally 2003; McCartney 2006; Thompson 1997, 2006). Nevertheless, even though the critical field is expanding there tends to be a lack of focus on the insecurities related to DNA evidence, both inside and outside the courtroom. As we are most often presented with the advantages and securities related to DNA evidence, this chapter will provide another side of the story: the *insecurities*. The advantages will be discussed only briefly to begin with. Consequently, the arguments presented will not be able to give an exhaustive analysis of DNA's qualities as a forensic tool. Rather the discussion will draw on the advantages and disadvantages of DNA evidence identified by Norwegian defence lawyers. The chapter draws on in-depth interviews with 15 prominent defence lawyers, selected on the grounds of their experiences with cases where DNA evidence played a significant role. Where relevant, quotes from interviews with expert witnesses on DNA are added. Nevertheless, the main focus is on the defence lawyers themselves. By its nature, the job of a defence lawyer is to point out weaknesses and insecurities of various types of evidence. Even so, the interviewed lawyers acknowledged that DNA evidence was beneficial not only for the prosecution but for their clients and for legal protection in general as well.

However, before discussing the lawyers' experiences with DNA, some background information is needed, especially for readers unfamiliar with the Norwegian court procedures. DNA evidence often requires the use of expert witnesses. Norwegian expert witnesses are appointed by the courts, a point which differs from several other countries, particularly in the Anglo-Saxon jurisdictions, where the parties themselves assign experts. Moreover, until recently there has existed only one forensic DNA laboratory in Norway. It is state-owned and functions as a unit within the University of Oslo. The police are required by law to use this public laboratory. There has been almost no tradition for using second opinion on forensic DNA evidence in Norwegian courts. However, in 2006 a new laboratory offering forensic DNA analyses was opened. So far, it has been used in only a handful of cases and simply to provide a second opinion. Consequently, some of the findings in this chapter may be applicable only to nations that share essential similarities with Norway – that knowledge of DNA evidence is more or less monopolised, limited to a small number of people and institutions, be that due to a formal legal monopoly or because the limited amount of forensic DNA-related work makes it impossible to have more than one DNA laboratory. However, other findings in the chapter may have a broader relevance for forensic DNA evidence.

A CRIMINAL CASE AS A JIGSAW PUZZLE AND DNA EVIDENCE AS ONE OF ITS PIECES

The chapter builds on the metaphor of a jigsaw puzzle as a means of exploring and critiquing the role of DNA in the courtroom. A jigsaw puzzle begins its existence as a clear, single picture. It is then cut into pieces and the pieces are mixed and scattered. Only when a substantial number of pieces have been gathered together, studied carefully and placed correctly relative to one another can we again see the original image clearly. This is also one way of seeing a criminal investigation ending in a court case. Bits of scattered evidence are gathered together, analysed in relation to one another and assembled into a picture of the original act, a picture that can convincingly be presented in court. In this view, DNA evidence is seen as a particularly crucial piece of the puzzle, rather like the straight edges, or corner pieces or, not least, the one piece of the puzzle that reveals the face of a portrait image. Despite this similarity there is one large difference between a criminal case and a puzzle. A jigsaw puzzle has only one solution, that is when all the pieces are in their right place and one can see the picture. A criminal case, however, may have several true pictures, depending on different points of view. People experience events differently. Two people may have had sexual relations; one participant believes it consensual, the other believes it was rape. As two different stories are told of the same act, pieces of evidence are presented to try to make visible the right picture. Furthermore, all the pieces in a criminal case jigsaw puzzle will never be available. A criminal case probably looks more like an old and well-used jigsaw puzzle, where some of the pieces are missing and others might have lost their original shape and hence do not fit as they should. Some of the pieces might have faded with time so that it will be difficult to envisage where in the picture they belong. Sometimes pieces from one jigsaw puzzle show up in the wrong box by accident. Other times pieces are put in the wrong box with the sole purpose of confusing and twisting the picture. Some pieces might even belong in more than one puzzle. The evidence that makes up the pieces in the jigsaw puzzle of a criminal case may be of a variety of forms ranging from confessions, witness testimonies, autopsies, fingerprints and DNA, to mention just some.

No piece in a criminal case jigsaw puzzle has a stable form or size; they are shaped by how they are presented, challenged and interpreted. Witness testimonies, for example, are pieces that are shaped by a complex interaction of perception, memory and social influence (Williamson 2007). This acknowledgement implies that courts cannot trust witness testimonies as much as they traditionally have (Johnsen 2007: 26). Confessions may also be erroneous, which increases the need to find expert evidence that can document the course of events without having to use witness testimonies, or that can enable a control mechanism of them (ibid.). However, lately also the accuracy of forensic sciences has been questioned. Saks and Koehler declare that '[c]onverging legal and scientific forces are pushing the traditional forensic identification sciences toward fundamental change' (Saks and Koehler 2005: 892). The assumption of uniqueness in several forensic sciences is weakened by evidence of errors in proficiency testing and in actual cases.

DNA is increasingly used and carries important weight as evidence in criminal cases (Kobilinsky et al 2005: 239). According to Imwinkelried (2004: 91), '[w]e are now well into the era of DNA evidence'. However, an important point is that a piece of DNA by itself does not make DNA evidence. DNA evidence is a construction. Several actions are required to transform a piece of DNA into DNA evidence which can then be used as a piece in a criminal case jigsaw puzzle. DNA has to be left somewhere, it has to be collected, it has to be processed, it has to be analysed, it has to be presented in a courtroom, and it has to be interpreted. This illustrates how DNA evidence is a construction that involves several 'construction workers', for example people committing an offence, police, DNA analysts, expert witnesses, lawyers (both prosecutors and defence), judges and jury members. All of these 'construction workers' may influence how the piece of jigsaw puzzle consisting of DNA evidence is shaped, presented and perceived in a trial. Despite this, DNA is often presented and perceived as an objective truth; a piece of jigsaw puzzle that has only one given size, shape and form. In the following, securities and insecurities that contribute to constructing DNA evidence will be discussed. However, the focus will be on the use of DNA evidence in court rather than the production of the evidence before it reaches the courtroom.

The securities of DNA evidence

In a court case it may be advantageous to have as much relevant information about the act which is up for trial as possible. Linking persons to a crime scene through forensic identification has been done for centuries. DNA technology contributes with an additional way of doing so. It is also considered '. . . the most systematic and rigorous way of individuating human beings and it is widely celebrated as a proven method for the reliable and repeatable identification of individuals and their bodily traces' (Johnson and Williams 2007: 39). One defence lawyer I interviewed said:

> Obviously DNA evidence is a piece of evidence that is very good as a starting point, because very few, nearly no people, except identical twins,

> have the same DNA. So clearly it makes identification possible of a person that has DNA in a certain place or on a particular object. It is nearly 100 per cent certain that one is identifying that the DNA comes from that person.[1]

DNA is also considered strong evidence by offenders. Having interviewed offenders about how they know and conceptualise DNA evidence, Prainsac and Kitzberger (forthcoming in Social Studies of Science 10) were told that if DNA is found in relation to a crime, offenders often see no point in denying the crime. When faced with DNA evidence, the interviewed offenders considered the game over, and felt that they might as well confess. An interviewed lawyer considered it a pro that DNA contributes to confessions as these can help to resolve unsolved crimes:

> I have often experienced that DNA evidence, and not least DNA registration, has contributed to unsolved cases being solved. That is very positive. These are very often cases where the perpetrator has been unknown.

By putting the question of guilt or innocence to rest, confessions also allow the defence lawyers to focus on other aspects of their clients' defence.

But because DNA is a probabilistic methodology, DNA is more effective in establishing non-identity, and (sometimes by necessity) innocence, than in establishing a suspect's identity, which is still not the same as establishing a suspect's guilt. Innocence projects have been established across the world, not least using forensic DNA technology. In the USA alone, DNA has enabled the exoneration of almost 200 inmates, several of them on death row (McCartney 2006: 186). In post-conviction exoneration cases forensic DNA technology may '. . . challenge the law's claim to being a truth-producing institution' (Cole 2007: 100). A person may be excluded as a suspect, or exonerated, if his DNA profile differs from the profile of a crime scene sample that is considered central to the case, for example semen at a rape scene. One of the lawyers I interviewed said the following about this:

> Very often it will help in an exclusion process. Look to the States and the DNA issue there. There are lots of cases where you have gotten a retrial because DNA has entered the field, and that means that it is an efficient tool to rule out somebody in relation to the individual criminal case and accusations. And naturally because of this nobody is actually against DNA as evidence.

Nevertheless, it should not be forgotten that overwhelmingly, DNA evidence is used to incriminate (Cole 2007: 100). This despite the fact that '[f]indings of

1 To protect the anonymity of my informants, I will not differentiate amongst them when publishing.

exclusion can generally justify greater statistical confidence and probative value than findings of inclusion' (Cole 2007: 100). DNA is more controversial as a tool of inclusion than of exclusion. The strength of inclusion depends, in part, on the numbers of DNA locations examined and the statistics reflecting how often the particular profile may be found in the general population. A DNA profile that occurs rarely in the population will more strongly suggest that the individual is the source of the biological evidence than would a common DNA profile (Turman 2001). However, the fact that there are massively more innocent than guilty among the population as a whole will always skew the predictive value of a positive identification downwards and the predictive value of a negative identification upwards. Exclusion will always be more certain than inclusion (Sætnan 2007). Increasing the number of DNA locations tested typically results in more powerful statistics, but often not all DNA locations are available from crime scene evidence (Turman 2001). The DNA profile may still be used to exclude if the DNA locations of the crime scene sample do not match with the locations of the profile of the suspect. Nevertheless there may be too few locations to use it as a tool of inclusion.

The insecurities of DNA evidence

Despite DNA evidence being presented as something that will increase justice and public security, there are clearly insecurities related to it, and all of the interviewed defence lawyers expressed disadvantages and concerns regarding the use of DNA as evidence. In the following some of the insecurities related to DNA as evidence, mentioned in the interviews, will be looked at in more detail. There was a general perception among the defence lawyers that:

> There is too little focus on the elements of uncertainty related to the use of DNA evidence.

Many of the interviewed defence lawyers claimed that this is especially worrying as DNA is a relatively new type of evidence. History has taught us that no forensic technique is foolproof or error-free (Cole 2004: 80; Saks and Koehler 2005: 892). Redmayne (2001) warns that even though DNA is in many ways beneficial for the criminal justice system (some of the benefits discussed above), we have to be careful to make sure that these benefits do not blind us so that we do not see the problematic issues that accompany DNA technology.

Forensic technologies have been blinding before. Little more than a decade ago forensic individualisation scientists compared pairs of marks, such as handwriting, fingerprints, tool marks, hair, tyre marks, bite marks, etc. Expert witnesses attested as to whether the marks matched, testifying in court that whoever or whatever made the one had made the other. The testimonies were rarely excluded or disregarded, and the foundations of the asserted expertise or the basis of the analyst's certainty were hardly ever questioned during cross-examination (Saks and Koehler 2005: 892). The evidence was treated as if it was 'black-boxed'. The concept of 'black-boxing' refers to the way scientific and technical work is made invisible by its own success. When something runs efficiently, or when a matter of fact is settled, the processes of interpretation and negotiation creating that 'fact' are often rendered invisible, leaving visible only inputs into and outputs of the processes, but none of their internal complexity (Latour 1999: 304). For example, shortly after fingerprints became legitimate markers of identity, it became nearly impossible to question forensic fingerprint judgements. Conclusions were treated as unassailable facts. It took nearly 100 years to acknowledge that fingerprints were fallible (Cole 2006). This highlights the great faith that existed both in expert witnesses and in forensic evidence.

Despite this great faith, that a number of forensic sciences drew on earlier, today's situation is different: 'Today, that once-complacent corner of the law and science interface has begun to unravel – or at least to regroup' (Saks and Koehler 2005: 892). Sceptics are attempting to open the black boxes surrounding a number of types of forensic evidence and in some cases have already succeeded. Scientists have started to query the core assumptions of a number of forensic sciences such as forensic identification of hair, fingerprints, bullets, footprints, handwriting and bite marks, mainly due to the reports of erroneous forensic identifications (ibid.). One of the components that enabled the processes of opening of the black boxes surrounding forensic evidence was the use of DNA evidence. In a number of cases DNA managed to exonerate suspects after they have been convicted and it became evident that the wrongful convictions were due to erroneous forensic expert testimonies. DNA also, to a much larger extent than many other forensic tools, provides a model for a scientifically sound identification science (ibid.). That the black boxes surrounding forensic evidence are opened implies that the insecurities surrounding the evidence are to a larger extent made visible. Ironically, however, when insecurities of forensic evidence are made visible, the evidence may contribute to increased security, as the assessment of evidence will be done on a more informed basis. These somewhat discredited forensic sciences may still be useful, once we have learned to deal with them with appropriate caution.

The next question we might then ask is whether there is a lesson to learn from this when dealing with new, and presumably stronger, more accurate forensic sciences. Even though DNA may challenge aspects of other forensic evidence, this does not mean that there are no insecurities related to the use of DNA itself. There needs to be an awareness of these insecurities, so that the black box surrounding DNA does not snap shut, as it did for so many other forensic sciences before – or perhaps more accurately, that we promptly reopen it once it snaps shut prematurely.

DNA AS SOMETHING MAGICAL

One of the main worries expressed by several of the defence lawyers is that there exist misconceptions about what DNA evidence actually represents and what it proves. They felt that there was a general perception in society of DNA evidence as something magical, a 'silver bullet' that will solve all crime. One defence lawyer said:

> I have the impression that people think it is almost something divine, an answer to the problem of evidence and doubt.

Another defence lawyer explained that DNA evidence is given too much weight due to the lack of focus on the insecurities:

> My experience is that DNA – what should I say? – it is given a bit too much weight. One doesn't see all the sources of errors.

Some also meant that there was a misconception of the strength and weaknesses of DNA as evidence in the criminal justice system. Not all parties in a case will have the same perception of how much weight DNA evidence should be given. Several of the defence lawyers I interviewed expressed a worry that DNA in general was given more weight than it should, that DNA was seen to represent more than what there is actually grounds for from the evidence. A defence lawyer said the following about assessing DNA evidence:

> I believe that one to a large degree overvalues what it is possible with certainty to derive from the evidence, and a worst case scenario then might be that we could get something that misguides us, rather than guides us about a fact. That is the main potential danger today.

Despite differences between the Norwegian and the Canadian criminal justice systems, research conducted in Canada found similar concerns among defence lawyers. Holmgren and Winterdyk (2001: 11) found that many Canadian lawyers felt vulnerable to DNA evidence and that they believed that 'DNA is so overwhelming that it is not possible to contest the evidence in courts'. It is difficult to challenge DNA in the courtroom because most people think it is virtually infallible (Thompson 2006: 15). If DNA is given too much weight and its power is overestimated then the use of DNA evidence may lead to wrongful convictions; hence we risk it being a tool that leads to injustice rather than justice.

When it is reported that DNA evidence is considered to represent more than what there is grounds for, this is mainly due to the inability to asses DNA evidence correctly, particularly due to the lack of knowledge among the parties involved. The lack of knowledge regarding DNA-related matters in the criminal justice system was one of the main worries amongst the interviewed defence lawyers. The defence lawyers expressed a concern about the knowledge level and the level of qualifications amongst all of the groups involved in constructing DNA evidence, such as the police, judges and DNA analysts. As one of the lawyers said:

> Perhaps we ascribe it greater evidentiary weight than there is factual foundation for. The involved participants understand to a very small extent what it really implies: judges, prosecutors, defence lawyers. They do not have enough knowledge to be able to ask the critical questions or check what the experts on forensic medicine have done.

However, the defence lawyers did not only express concerns regarding the knowledge level of the other involved parties; they also expressed uneasiness regarding their colleagues' knowledge level and, not least, their own. As one lawyer honestly admitted:

> I don't actually think that I have the skills to be able to plead that a piece of DNA evidence isn't valid.

Furthermore, not only the defence lawyers expressed concerns about the lack of knowledge of DNA-related matters amongst the involved parties. The following quote from an interviewed expert witness reveals a similar concern:

> I have very mixed feelings when I walk out of courtrooms. Of course I try to see if people are following what I am saying. But one thing is to understand the words; another thing is to understand the meaning and how to use it afterwards. We have seen examples that they misunderstand. They do not have the knowledge to use the information they are given. So what they have grasped and used appears as meaningless.

As defence lawyers' questioning of expert witnesses contributes to how evidence is perceived by judges and juries, their questioning of evidence influences to a large degree what aspects of the evidence come across in court rooms. It is easier to know which questions to ask, and what questions and answers are relevant for a case, if one has thorough knowledge of the topic. Consequently lack of knowledge or understanding amongst the defence lawyers may contribute to evidence not being adequately challenged, which in turn may lead to aspects of the evidence not coming across to judges and juries. The following quote from an interviewed expert witness reveals the limited questioning of DNA evidence in Norwegian courts:

> When you don't know something, then you accept what is being said, and then you don't ask any questions; no control questions, no critical

> questions. You don't dig deeper into the underlying question. You don't see the value of everything that is being said.

The reported lack of knowledge amongst the involved parties may be a reason for great concern. If the knowledge level is not adequate to challenge the presented evidence, then a piece of evidence cannot be given the right shape and place it should have had in a case. According to McCartney (2006: xx), this might have fatal consequences as '. . . miscarriages of justice will flourish in a culture which fails to properly scrutinise and question "scientific" evidence'.

However, inadequate knowledge of forensics cannot be eliminated if funding to educate the defence bar is not provided (Berger 2004: 121). The defence lawyers I interviewed expressed a need for more knowledge and a better understanding of DNA to be able to properly challenge DNA evidence, which would enable better defence for clients. These findings have evident similarities with what Holmgren and Winterdyk (2001) found when interviewing Canadian defence lawyers. Both Norwegian and Canadian defence lawyers also said that an adequate level of expertise about DNA was hard to achieve due to busy schedules and other priorities. Most of the defence lawyers I interviewed said they were self-taught on the topic of DNA. Some of them had read books about the subject in English, while others searched the internet for information. A few had followed a course on the topic which had lasted a couple of hours. At the time, this course was taught by the representatives from the Norwegian governmental DNA laboratory. Hence the course was taught by the same institutions that supply DNA analysis for Norwegian courtrooms and that supply expert witnesses on DNA in Norwegian courtrooms. Perhaps even more importantly, the defence lawyers pointed out that achieving knowledge about DNA was a learning-by-doing process. The next quote shows an example of a client's dependence on his or her defence lawyer's level of knowledge about DNA evidence:

> The point was that the police misread the DNA analysis, and the public prosecutor dropped the charges when we were in court when he saw the small comment that this could just as well be skin cells as something else. And that it actually probably was skin cells. Then their entire case lost validity. A man didn't have to go to jail for four years because they had misread a DNA analysis. This is an example that one has too much faith in DNA as evidence.

Skin cells spread more readily than, say, spittle, semen or blood, and skin cells could have ended up at the crime scene in variety of ways. Hence the DNA evidence could not prove what the public prosecutor had believed it could prove. The defence lawyer was experienced and was able to see this relevant nuance for his client. Yet the police had not been able to see the same relevant nuance, and a defence lawyer still in his learning phase would probably not have managed to either. This indicates the potential danger of having inexperienced lawyers who are at the beginning of their learning-by-doing-process. The example also illustrates the next point I shall discuss in this chapter: the strong position that DNA has as evidence.

When DNA evidence is the only evidence

One piece of jigsaw puzzle can never make up a jigsaw puzzle on its own, just as only one piece of evidence is not enough to make up a criminal case on its own. DNA evidence is no exception. DNA evidence can only be circumstantial evidence which cannot be directly linked to an action, but can shed light on aspects of it (Strandbakken 2003: 46). DNA evidence can only show a relationship between a person and an object, but cannot say anything about how it got there, or when. A defence lawyer explained it like this:

> Clearly a piece of DNA evidence cannot be used alone because alone it can't say much about the possibility that a person has committed a punishable act. It has to be supplemented with other things.

Despite this, several of the defence lawyers worried that DNA evidence was not used accordingly. They criticised the public prosecutors for pressing charges when their only evidence was DNA. As one defence lawyer expressed this worry:

> I think the prosecution often is pretty uncritical and just prosecute on the basis of DNA. And it is clearly a problem because when the prosecution has raised charges the possibility for conviction is very often pretty large. When charges are made, they have considered the case and even though one is supposed to take it from the beginning, it is not to be ignored that many are influenced by the fact that charges had been made. And I am not only thinking about lay judges, but actually also about professional judges.

The quote illustrates the great faith in DNA, and how a type of evidence that is only supposed to be a piece in the jigsaw puzzle actually may on its own become the entire picture. 'Because of its apparent objectivity, genetic information has become more than a source of evidence – it is becoming a form of proof' (Gerlach 2004: 89). If it is the case that suspects are taken to court on hardly any other grounds than DNA, then there is a need for concern, particularly when one considers the weight that DNA carries as evidence and the general lack of knowledge about it.

Dependence on the expert witness

Another concern expressed by the interviewed defence lawyers was that due to the lack of knowledge about DNA, they felt very dependent on expert witnesses. In the following quote from a group interview with one DNA expert witness and one DNA laboratory worker it becomes apparent that there are cases where not all relevant aspects of DNA evidence are mentioned or considered in the courtroom:

> Expert witness: In many cases I just come and say what I am to say. I do it my way, no questions, and I leave.
> DNA lab worker: And you can't tell the court that this is what you should ask. It's not one's task. So if they don't ask then . . .

It seems as if DNA has such a powerful effect that some of the participants seem blinded by it and hence do not question it thoroughly. The lack of knowledge, reported both by defence lawyers themselves and the expert witnesses, may hinder defence lawyers from challenging the evidence since they do not feel confident to ask the relevant questions that might reopen the black box of DNA technology. Another reason for the black box not being adequately opened may also be a misconception about how it should be opened. Defence lawyers did not express that they considered it the responsibility of the expert witness to open the black box, however, they did express a dependency on the expert witness, a sort of faith that the expert witness would know what to say when the defence lawyers' knowledge was not sufficient. The expert witness, however, does not consider it her task to open the black box. It would also be difficult for an expert witness to do this as expert witnesses do not have thorough knowledge about the specific cases and hence do not always know the relevance of the evidence that they are there to present.

In the following quote an interviewed lawyer synthesises his thoughts about the inability of the defence to challenge DNA expertise:

> Clearly we aren't well enough trained to go to the battle against the experts in forensic medicine, and we have only one institute in Norway that does DNA analyses, at least it's been like that until now. And it is difficult to obtain other viewpoints, thoughts and expertise than what one gets from the government.

Another lawyer commented on the Norwegian system in the following way, thereby giving a glimpse of insecurity related to DNA evidence:

> It is an obvious shortcoming; we have a system that is based on an ultimate truth, and that is the result. And because the perception is that

> there is only one answer, there is no point sending it [the DNA sample] to two places. As long as this is the perception it will remain like that.

Having only one provider of DNA analysis and only one provider of expert witnesses results in there being only one provider of 'truth' about forensic DNA evidence. As a consequence, there is only one 'truth' provided in the Norwegian courts about a type of evidence which is given much weight and considered objective by several of the involved parties. This monopoly on 'truth', combined with the faith in the objectivity of that 'truth', is alarming to those who see that 'truth' as being less than absolutely certain and as a result of subjective interpretations. The forensic DNA situation becomes vulnerable when there is only one provider of the forensic DNA evidence 'truth' and lawyers do not always feel they have adequate knowledge to question the evidence properly.

The vulnerability of the Norwegian DNA system becomes especially visible keeping in mind how defence lawyers reported feeling dependent on the expert witnesses, also evident in the following quote from one of the lawyers:

> The problem is that in real life, it is mostly one person in Norway who attends all court cases and tells about DNA. This doesn't lead to very much professional debate.

According to Cole (2004: 80), '. . . we should be concerned about allowing law enforcement to monopolize expertise in the area of forensic DNA typing'. Recent history has demonstrated that DNA evidence has been subject to conscious and unconscious pro-prosecution bias (Cole 2004: 80 and Thompson et al 2003). One of the interviewed defence lawyers pointed out the problem:

> They [the Norwegian DNA lab] work for the public prosecutor and yet appear as neutral for the court of justice, and that is a problem. There is nobody that you as a defence lawyer may turn to to get information or send something to.

Several of the defence lawyers also expressed the need to have a possibility of asking for a second opinion from an independent body, but until recently this has not been possible in Norway. When asked why they do not choose to go abroad with the DNA evidence a defence lawyer answered:

> It has to do with the traditions of the Norwegian judicial system. Most suspects can't afford to hire experts from abroad. And defence lawyers aren't paid to hire them . . . the courts believe that we have someone in the country who can tell us this. We don't need another one. When the court has appointed one expert witness that should be enough. In that way it is hard to challenge the problem.

Taking this to the furthest consequence it implies that if a defendant is involved in a case where DNA plays a large role, the legal protection of the individual in Norway depends on whether the defendant has got the financial means or not to fund an independent second opinion. Jasanoff (2006: 334) writes of the practice as follows: 'Indigent defendants, who cannot afford effective lawyering, may find their fates decided less by the strength of the scientific evidence as assessed by technical experts than by the vigour and ingenuity of the advocacy mobilized by their defence.'

There is the possibility of getting the state to pay for a second opinion, but then the defence lawyers have to convince a judge that it is required. As Holmgren and Winterdyk (2001: 12) point out, to be able to make the state pay for a second opinion requires a certain amount of knowledge, which as we saw above, not all of the defence lawyers possess. Often the decision as to whether the state should pay for the second expert witness is not made before the case reaches the court; this implies that the defendant has to take the risk of not getting his expenses refunded. Having to go abroad for a second opinion may not be only a financial challenge but may also provide another problem, namely the language barrier. DNA in itself is seen as challenging evidence to understand for judges and juries. Expert witnesses are often perceived as talking another language, because their fields are so complicated; if the expert witness also literarily talks another language, this might result in double trouble. Not only will the listeners have to hear about a difficult piece of evidence to start with, but they might have to hear it explained in a language that is not their mother tongue.

When the wrong piece fits perfectly – by accident

When laying a jigsaw puzzle a piece may appear to fit well in one position, yet not actually be in its right place. That a piece is misplaced might not be revealed until the rest of the pieces fall into place. In a criminal case all the jigsaw puzzle pieces will never be available, so when a piece seems to fit into what one thinks is the picture of the crime, that piece might become conclusive evidence, despite being in the wrong place. One of the interviewed defence lawyers used the following example from a famous Norwegian case to illustrate how DNA has the possibility to become conclusive evidence or the final proof, even when it should not be. A policeman was shot during a robbery by one of 13 robbers. A witness held hostage by the man who shot testified that the man who shot was white. When the murder weapon was found and analysed for DNA there was DNA on the weapon: DNA from the only defendant with dark skin. The defence lawyer went on to explain what he thought would have happened if the DNA found on the murder weapon had matched the suspected white person, even when other evidence pointing to that suspect was inconclusive:

> If one is left with a kind of nagging feeling that it is probably him that did it and in addition we have got his DNA on the murder weapon. Yes! Then it is seen as conclusive evidence. And then you see how something that is actually just corroborative evidence becomes cardinal evidence. And that is dangerous, very dangerous in the judiciary! There isn't focus enough on the alternatives.

Incriminating others

We all shed DNA every day: when we spit, when our hair and skin cells fall off, etc. This makes DNA traces a double-edged sword, the positive edge being that people committing crime easily leave DNA at a crime scene, the negative edge being that it may be easily planted by others. Several defence lawyers thus reported stories told by their clients about how DNA had been left at a crime scene in order to incriminate innocent people and to confuse investigators:

> I have heard from criminals that if they are planning a break-in for example, or something like that, or to do something serious, then they usually drop by a disco first and gather cigarette butts, to leave some at the crime scene of the break-in. Their objective is to confuse the police and first of all they will hope that it will become evidence of exclusion, right? It is an unknown perpetrator here, and it is not me. In addition another signal of danger in relation to that kind of use is that if you start to plant cigarette butts or other things from labelled people or by a coincidence a person who is on the DNA database that has been to the disco, then he can start having a problem, right? Partly because you can plant things directly and partly because you can make sure innocent people are suspected because one believes too much in the DNA evidence.

At a first glance the thought of planting DNA evidence might appear more relevant for police investigations than during a trial, but according to the defence lawyer, the planting of DNA evidence may be done with the goal of framing other criminals – people who already have a criminal record, people who are on the DNA database, people who may wind up as suspects in court with the DNA evidence and their prior records are being used against them:

> If you are sitting with a client who has a long criminal record it is extremely difficult to be able to get someone to believe that the suspect's DNA has intentionally been planted by other criminals. It is so easy for a prosecutor to say: 'Of course. Here it comes. I have just been sitting here waiting for it. Here comes that kind of explanation.' Then the prosecutor can ridicule it. And with some luck he will get the court's acceptance for it.

Since the interviewed defence lawyers have been told that planting of DNA is happening, they worry about the time they will have to take such a case to court:

> This might be a good example of how dangerous DNA may be as evidence; I do believe it would be very difficult to convince a jury and a professional judge that it is planted. I think it is nearly impossible! It makes statements very vulnerable to be attacked as constructed. I think we will have some very scary cases like that coming up. I think we have only seen the beginning of it. The paradox then is that DNA may become an enemy of the rule of law, and not the opposite.

CONCLUSION

In a jigsaw puzzle certain pieces are more important than others. For a piece of a jigsaw puzzle to contribute in a positive manner it has to be used with caution. If new pieces, which are blown out of proportion, are added, or if they are mistakenly perceived as more important than they actually are, or even worse do not belong to that particular jigsaw puzzle at all, then they might be of more harm than help. Despite its increasing use in the last 20 years, DNA is still a relatively new type of evidence. It is often presented quite one-dimensionally, simply as a security-increasing measure, and there is little doubt about DNA's usefulness in criminal cases. When 15 of the most prominent defence lawyers in Norway were interviewed about DNA they all praised it as a useful new type of forensic evidence, particularly since it can also lead to exoneration of the innocent. Even so, this chapter has shown that the interviewed lawyers had a number of concerns regarding DNA. They were worried that the power of DNA is overestimated, that charges were pressed when DNA was the only evidence, that planting of DNA evidence confuses not only the police but also judges and juries. Not least, they were very concerned about the lack of focus on insecurities related to forensic DNA evidence, both inside and outside the courtroom.

It appears as if there is a black box surrounding DNA evidence which makes the insecurities related to DNA evidence invisible. There may be several reasons for this. The lawyers mentioned that they were concerned about the apparent lack of knowledge regarding DNA evidence among the involved parties, such as the police, prosecutors, judges, jury members, and not least themselves. It is easier to question something one has a thorough understanding of; therefore this lack of understanding of DNA evidence results in lawyers not always knowing how to question DNA evidence. Correspondingly, expert witnesses can then pull up evidence from that black box like rabbits in a magic trick. This evidence is then not challenged adequately and therefore not given its proper shape or place in the puzzle. If the evidence is not adequately questioned there is a risk that not all relevant aspects come across in the courtroom. The end result may then be judicial errors, sometimes even miscarriages of justice. In other words, DNA evidence may lead not only to increased justice and increased security but also to injustice and insecurity.

No science or technology is all good or all bad, and all sciences and technologies are inherently risky (Collins and Pinch 1994: 150). In contemporary society a number of technologies are used to increase security – e.g. CCTV cameras, and a wide range of biometric technology such as gait recognition, facial recognition, iris scans and voice recognition. Results of these technologies are also finding their ways into the courtrooms as evidence. Despite often being presented as one-dimensional, as all good and simply as technologies contributing to increased security, there are insecurities related to these technologies. However, there tends to be little focus on these insecurities both inside and outside the courtroom. This is unfortunate as these technologies may be, if properly applied, able to contribute to increased security and increased levels of justice. The purpose of this chapter has been to point out that the more knowledge we have of these technologies, and of the securities and the insecurities they entail, the better our premises for using them correctly. Therefore, in the long run, no one is well served by the one-dimensional picture of these technologies and by the black-boxing of the insecurities related to them. If the insecurities remain hidden in the black boxes there is a risk they will not simply lead to enhanced security but will work against their intentions and lead to increased insecurity.

References

Berger, M.A. (2004) 'Lessons from DNA: Restricting the balance between finality and justice', in D. Lazer (ed) *DNA and the Criminal Justice System: The technology of justice*, Massachusetts: MIT Press, 109–132.

Cole, S. (2007) 'How much justice can technology afford? The impact of DNA technology on equal criminal justice', *Science & Public Policy*, 34(2): 95–107.

Cole, S. (2006) 'The myth of fingerprints: The legacy of forensic fingerprinting and arrestee databases', *GeneWatch*, 19(6): 3–6.

Cole, S. (2004) 'Fingerprint identification and the criminal justice system', in D. Lazer (ed) *DNA and the Criminal Justice System: The technology of justice*, Massachusetts: MIT Press, 63–91.

Collins, H. and Pinch, T. (1994) *The Golem: What everyone should know about science*, Cambridge: Cambridge University Press.

Dagsavisen (2007) 'Nye milliarder til justissektoren', 24 July.

Diesen, C. and Björkman, J. (2003) 'DNA-bevis är inte alltid starka', *Juridisk Tidsskrift*, 4(4).

Gerlach, N. (2004) *The Genetic Imaginary: DNA in the Canadian criminal justice system*, Toronto: University of Toronto Press Incorporated.

Holmgren, J. (2003) *Beyond the Walls of the Laboratory: judge and jury interpretations, perceptions, and understanding of DNA evidence*, Calgary: University of Calgary.

Holmgren, J. and Winterdyk, J. (2001) 'DNA evidence: Balancing the scales of justice', *LawNow*, 26(2): 11–13.

Home Office (2004) *Building Communities, Beating Crime: A better police service for the 21st century*, London: Home Office.

Imwinkelried, E.J. (2004) 'Fingerprint identification and the criminal justice system: Historical lessons for the DNA debate', in D. Lazer (ed) *DNA and the Criminal Justice System: The technology of justice*, Massachusetts: MIT Press.

Jasanoff, S. (2006) 'Just evidence: The limits of science in the legal process', *The Journal of Law, Medicine and Ethics*, 34(2): 328–341.

Johnsen, J.T. (2007) 'Feilkilder ved ekspertbevis. Hvordan kan de påvirke utfallet av straffesaker?', in P. Brandtzæg and S. Eskeland (eds) *Rettsmedisinsk sakkyndighet i fortid, nåtid og fremtid*, Oslo: Cappelen.

Johnson, P. and Williams, R. (2007) 'European securitisation and biometric identification: The uses of genetic profiling', *Annals of the Italian National Institute of Health*, 43(1): 36–43.

Kobilinsky, L., Lotti, T.F. and Oeser-Sweat, J. (2005) *DNA Forensic and Legal Applications*, New Jersey: Wiley-Interscience.

Koehler, J.J. (2001) 'The psychology of numbers in the courtroom: How to make DNA match statistics seem impressive or insufficient', *Southern California Law Review*, 74: 1275–1306.

Latour, B. (1999) *Pandora's Hope: Essays on the reality of science studies*, London: Harvard University Press.

Lazer, D. (ed) (2004) *DNA and the Criminal Justice System: The technology of justice*, Massachusetts: MIT Press.

Lynch, M. and McNally, R. (2003) ' "Science," "common sense," and DNA evidence: A legal controversy about the public understanding of science', *Public Understanding of Science*, 12: 83–103.

McCartney, C. (2006) *Forensic Identification and Criminal Justice: Forensic science, justice and risk*, Cullompton: Willan Publishing.

Ministry of Justice and the Police (2007) 'Pressemelding [press release] Auka bruk av DNA for å oppklare meir', available at http://www.regjeringen.no/nb/dep/jd/pressesenter/pressemeldinger/2007/Auka-bruk-av-DNA-for-a-oppklare-meir.html?id=482547 (accessed 25 January 2008).

Prainsac, B. and Kitzberger, M. (forthcoming in Social Studies of Science) 'DNA behind bars: "Other" ways of knowing forensic DNA technologies'.

Redmayne, M. (2001) *Expert Evidence and Criminal Justice*, Oxford: Oxford University Press.

Sætnan, A.R. (2007) 'Nothing to hide, nothing to fear?: Assessing technologies for diagnosis of security risks', *International Criminal Justice Review*, 17(9): 193–206.

Saks, M.J. and Koehler, J.J. (2005) 'The coming paradigm shift in forensic identification science', *Science*, 309: 892–895.

Storberget, K. (2007) 'Vi skal oppklare mer', *Østlendingen*, 23 October.

Strandbakken, A. (2003) *Uskyldspresumpsjonen – In dubio pro reo*, Bergen: Fagbokforlaget Vigmostad & Bjørke AS.

Thompson, W.C. (2006) 'Tarnish on the "gold standard": understanding recent problems in forensic DNA testing', *The Champion*, (10): 14–20.

Thompson, W.C. (1997) 'A sociological perspective on the science of forensic DNA testing', *U.C. Davis Law Review*, 30(4): 1113–1136.

Thompson, W.C., Taroni, F. and Aitken, C.G.G. (2003) 'How the probability of a false positive affects the value of DNA evidence', *Journal of Forensic Sciences*, 48(1): 47–54.

Turman, K.M. (2001) 'Understanding DNA evidence: A guide for victim service providers', available at http://www.ojp.usdoj.gov/ovc/publications/bulletins/dna_4_2001/welcome.html (accessed 8 December 2007).

Williamson, T. (2007) 'Psychology and criminal investigation', in T. Newburn, T. Williamson and A. Wright (eds) *Handbook of Criminal Investigation*, Cullompton: Willan Publishing.

Chapter 12

'Catastrophic moral horror'

Torture, terror and rights

Vidar Halvorsen

When the notorious photos from Iraq's Abu Ghraib prison were broadcast worldwide in April 2004, the official US response invoked the well-known theory of a few rotten apples in an otherwise untainted barrel of professional detention and interrogation practices.

For historian Alfred McCoy, however, the photo of a hooded Iraqi, standing in an up-right position on a box with fake electrical wires hanging from his extended arms, illustrated two essential ingredients in a historically well-established pattern of so-called psychological interrogation techniques. The hood generates sensory disorientation or deprivation; the extension of arms is a stress position technique for the self-infliction of pain. Furthermore, the photo of Private Lynndie England, holding an Iraqi detainee in a leash like a dog, provides another illustration of a closely related technique, namely, the systematic attempt to undermine cultural identity by way of sexual humiliation and by exploiting an individual's fears and phobias.

As McCoy convincingly demonstrates in *A Question of Torture* (2006), these techniques were based on extensive psychological research financed by the CIA and the American military during the 1950s and the beginning of the 1960s and subsequently implemented primarily in Asia and Latin America. I mention the historical background of these interrogation practices not merely because it serves to make sense of their reappearance after September 11, but also because it might serve to make sense of an emerging, public debate or discourse on the meaning of torture and its possible justifiability.

One might object that there is nothing really new in these debates on the justifiability of torture or torture-like interrogation techniques; that they merely provide yet another illustration of what is commonly described as a 'balancing' of the vital moral goods of security versus liberty. Scholars and practitioners of criminal justice, for example, are already familiar with Herbert Packer's (1968) influential distinction between 'due process' and 'crime control'. However, from being a useful device of moral reasoning in theoretical ethics, thought-experiments involving so-called 'ticking bomb scenarios' are now increasingly invoked by politicians and media commentators to suggest that the threat of terrorism has drastically altered the stakes, thus requiring an equally drastic reconsideration of the means necessary to rebalance the scales of liberty and security. Moreover, such justificatory attempts resonate with images of heroic responses to moral horror in popular culture, as embodied by Clint Eastwood in *Dirty Harry* and Kiefer Sutherland in *24*. 'If the ends don't justify the means, what does?' is reported to have been the favourite aphorism of Robert Moses, who vigorously defended the instrumental rationality of post-war urban planning in New York (Caro 1974). Yet, from the fact that noble ends are essential ingredients in the justification of means, it does not follow that we are entitled to ignore moral side-constraints on their use.

THE PROHIBITION AGAINST TORTURE

Although the prohibition against torture is the only *absolute* human right, well-known initiatives have ignited controversies concerning its meaning and scope. For example, on 19 January 2002, Defence Secretary Donald Rumsfeld declared that the Geneva Conventions of 1949, which protect prisoners of war against the infliction of torture and inhuman treatment, did not apply to 'unlawful combatants' captured in Afghanistan or elsewhere and subsequently transported to Guantanamo Bay and other detention facilities around the globe. In August 2002, Assistant Attorney General Jay Bybee issued the now infamous 'torture memo', in which it was suggested that torture should be conceived of as pain associated with 'serious physical injury so severe that death, organ failure, or permanent damage' results.

In making the case for a distinction between a physicalist conception of unacceptable torture techniques on the one hand and a psychological conception of acceptable interrogation practices on the other, Bybee and his associates made an interesting reference to a European court case. In *Ireland v The United Kingdom* (1978), the European Court of Human Rights (ECHR) ruled that the widespread and systematic use of five interrogation techniques against detainees suspected of IRA terrorism in the 1970s – wall-standing, hooding, exposition to noise, and deprivation of food/water and sleep – did *not* constitute torture in terms of article 3 in the European Convention on Human Rights. Crucially, however, Bybee missed an important point in his memo, namely the point that according to the court, there had nevertheless been a violation of article 3, which states: 'No one shall be subjected to torture or to inhuman or degrading treatment or punishment.'

Bybee was right to insist that on the court's own explicit account, the basis for article 3's distinction between torture, inhuman and degrading treatment or punishment is the severity of pain and injury suffered by victims. The term 'torture', then, is, as in Bybee's memo, reserved for practices falling on the upper end of a scale of harmful consequences. As the court has made clear in a variety of cases, practices that fail to reach the threshold of torture may nevertheless constitute a violation of article 3's prohibition against inhuman or degrading treatment or punishment. Yet this threshold conception of torture does not exhaust the court's understanding of the nature of torture, as Malcolm Evans and Rod Morgan make convincingly clear in their impressive work on *Preventing Torture* (1998). In the court's adjudicative practice, two further elements are operative: the underlying purpose of torture and the context of its infliction.

The basic purpose of torture is the obtaining of information and confessions or the imposition of punishment, or both. The paradigmatic context of torture is a situation in which the victim is already under complete physical control by his torturers and vulnerably exposed to their powers. Thus, although cases involving intentional infliction of pain and injury might be indistinguishable in terms of the threshold approach, they might be distinguishable in terms of the contextual and purposive elements invoked by the European Court of Human Rights.

Indeed, the contextual ingredient has a profound bearing on the question of justification. Reasonable people might disagree on whether, say, the shooting by the police of a dangerous, fleeing criminal is legitimate or illegitimate. However, reasonable people are reasonable precisely by virtue of their willingness to provide arguments to vindicate their conclusions in public debates. In a variety of cases involving coercive force and violence by the police, answers to questions of legitimacy are sometimes not easily available, and moral deliberation seems inescapable.

With torture, inhuman and degrading treatment or punishment, the situation is radically different. In the eyes of the ECHR, the status of article 3 is absolute and non-derogatory; from the moment a given practice has been subsumed under the article, no question arises as to its legitimacy. This is a 'default position', the validity of which has been increasingly challenged in public debate after September 11. If we accept it, however, as I think we should, it follows that many of the cases analysed in empirical research on police violence are *ethically* trivial in the sense that no moral deliberation is needed to establish their illegitimacy.

Ethics is a subject reflecting on the morality (and immorality) of human practices. Consider, for example, the notorious Louima case: on 9 August 1997, in the restroom of the 70th precinct station house in Brooklyn, NYPD officer Justin A. Volpe shoved a wooden stick into the rectum of Haitian immigrant Abner Louima, who suffered a torn rectum and ruptured bladder that hospitalised him for two months. Needless to say, such injuries are hard to rationalise in terms of the usual 'cover charges' of disorderly conduct, resisting arrest and assault. Instead, Volpe's defence attorney, Marvyn Kornberg, argued that Louima had suffered the injuries in an act of consensual, homosexual intercourse prior to his arrest. Surprisingly and quite atypically, however, on 8 June 1999 Volpe pleaded guilty to charges of aggravated sexual abuse and first-degree assault and was later sentenced to 30 years in prison. The 'default position' implies that there can be no ethically interesting discourse on the legitimacy of non-consensual police sodomy. Characteristically, the lack of valid reasons for these kinds of police abuses is manifest in the more or less routine denials of their existence or in the legal 'cover charges' rationalising their occurrence.

The Dirty Harry problem

The 'default position' has, as already mentioned, been challenged, first most notably by the Israeli Landau Commission, which was established in 1987 to investigate the lawfulness of interrogation methods employed by the General Security Service (GSS) against Palestinians suspected of terrorism.[1] Invoking the well-known 'ticking bomb scenario', the Commission argued that techniques involving 'physical pressure', the exact nature of which had to be kept secret to retain their effectiveness, were justified on grounds of their necessity for the prevention of a greater evil. 'The ticking bomb scenario' refers to a situation in which a detained adversary, known to have programmed a lethal device to explode, say, in a crowded shopping centre in the very near future, is interrogated about the exact location of the device. Could we justifiably inflict pain and injury on the suspect to obtain the vital, life-saving information? 'The answer,' the Landau Commission concluded, 'is self-evident.'

However, the Commission was unable to identify GSS cases that could plausibly be said to illustrate the real-life relevance of 'the ticking bomb scenario' and thus provide a justification for the agency's systematic use of physical interrogation techniques against terrorist suspects. In police ethics, the usual starting point for discussion is Carl Klockars' classic article on 'The Dirty Harry Problem' (1985). Klockars' description of the problem is not informed by empirical works on policing but draws on Don Siegel's Hollywood movie, *Dirty Harry* (1971), starring Clint Eastwood in the role of Detective 'Dirty Harry' Callahan. A psychopathic serial killer, Scorpio, has kidnapped a 14-year-old girl and holds her captive under conditions with so little oxygen available that she will die in a few hours unless a $200,000 ransom is paid by the city of San Francisco. Callahan delivers the money to Scorpio, who refuses to act on his promise to release the girl. After a series of dramatic events, Callahan confronts Scorpio on the playing field of a football stadium and shoots him in the leg. On the belief that there is a possibility that the girl might be alive, Callahan twists his foot on Scorpio's wounded leg to obtain the necessary information concerning the girl's location. Alas, it turns out that the girl is dead. Was 'Dirty' Harry nevertheless justified in the application of 'dirty means' in his attempt to achieve a noble end? The movie does

1 See the special edition of the *Israel Law Review* (vol. 23, no. 2–3, 1989), which includes excerpts from the commission's report.

not provide the audience with much doubt about what the correct answer must be. Furthermore, an overwhelming majority of recruits regularly reach the same conclusion as the Landau Commission when challenged to respond to various versions of the 'Dirty Harry' dilemma in ethics classes at police academies around the globe.

Unlike methods of police interrogation, which are typically employed to obtain evidence of a suspect's *past* wrongdoing, 'the ticking bomb scenario' and the 'Dirty Harry problem' are now, after September 11, seen as instances of the right of self-defence and defence of others (who are unable to defend themselves): these cases raise the question of what we are morally permitted to do to ward off an *imminent* and unjust lethal attack. Although the perpetrator's body has already been physically constrained, as in cases of torture, the ticking bomb is still beyond our control, thus posing a serious threat to the lives of innocent people. On the right of self-defence, which reflects a strong moral intuition shared by most people, we are morally entitled to shoot and seriously wound or kill a person who is caught in the process of detonating a bomb. This requirement of imminence is at least partially preserved in some real-life cases of modern warfare, as when Israel launched a first strike on Egypt in The Six Day War of 1967. In contrast, it was strikingly absent in one of President Bush's first formulations (on 1 June 2002) of the new security doctrine of pre-emption:

> Yet the war on terror will not be won on the defensive. We must take the battle to the enemy, disrupt his plans, and confront the worst threats before they emerge. In the world we have entered, the only path to safety is the path of action. And this nation will act.
> (Graduation Speech at West Point, available at the White House website)

The fatal consequences of a more comprehensive logic of pre-emption, with its relaxation of the basic requirement of epistemic justification underlying the principle of imminence, have been unfolding since 20 March 2003, when USA and its 'coalition of the willing' invaded Iraq.

A much more plausible version of the principle of pre-emption was applied by the Special Air Service (SAS) in the ECHR case of *McCann and Others v the United Kingdom* (1995). Prior intelligence indicated that the Provisional IRA was planning a terrorist attack on Gibraltar in March 1988. Moreover, use of a remote-control device to detonate a car bomb was considered to be the modus operandi most likely to be used in the attack. An extensive surveillance operation enabled local police to identify three IRA members, and four plain-clothes soldiers from the SAS surveillance team followed the suspects in the streets of Gibraltar.

Acting on the belief that the suspects were reaching for the button of a remote-control device, the soldiers fired altogether 27 rounds, killing the suspects immediately. Subsequent searches established that the deceased were unarmed and that their car did not contain a bomb, although a second car was discovered two days later, containing explosives and timing devices.

The Strasbourg court found that the soldiers' descriptions were substantiated by independent eyewitness accounts and that their split-second interpretations of the suspects' movements were reasonable in the circumstances of the case. Nevertheless, although the killings, taken in isolation, did not constitute a violation of the right to life guaranteed by article 2 in the European convention, the court's majority (10–9) ruled that the anti-terrorist operation as a whole had been controlled and planned in a manner which failed to respect the requirements of article 2.

The crucial question is whether the self-defence justification of intentional injuring or killing in 'the remote-control scenario' can be extended to cover the intentional infliction of pain and injury in 'the ticking bomb scenario'. There is a difference between the two scenarios: in 'the remote-control case' but not in 'the ticking bomb case', we operate on the person's body to counteract and neutralise his ongoing, lethal actions; in 'the ticking bomb case' but not in 'the remote-control case', we operate on a person's mind by way of inflicting (coercive) pain and injury on a body which has already been neutralised. Now, is this difference a morally relevant difference?

The ECHR has argued that the distinction between body and mind is legally and morally relevant as far as the acceptability of evidence is concerned. For example, on the court's account, the right to remain silent 'does not extend to the use in criminal proceedings of material which may be obtained from the accused through the use of compulsory powers but which has an existence independent of the will of the suspect'. The court's argument explains why coercion necessarily must nullify the evidential status of confessions, since such evidence, unlike objects and facts, does not have an existence independent of a person's will or, more generally, of his rationality.

Confessions are relevant to the extent that they serve to establish a justificatory, evidential connection between certain facts on the one hand and the truth of a specific proposition on the other. Because confessions cannot exist independently of suspects' rationality, the normal legal effects of confessions must be nullified to the extent that they have been obtained through the infliction of torture or the threats of torture, which are paradigmatic instances of coerced confessions. A torture-based confession is the outcome of an essentially unreasonable choice in the sense that it replaces a justificatory relation between facts and beliefs with a causal or quasi-causal relation generated by pain or fear of pain. The replacement of justification with causation explains, of course, the notorious and well-known unreliability of torture-based interrogation techniques.

'CATASTROPHIC MORAL HORROR'

However, the obtaining of *evidence* for past wrongdoing lacks the urgency of a situation in which we are in desperate need of *information* that may enable us to ward off, in Robert Nozick's terminology, 'catastrophic moral horror' (1974: 30) in the very near future. Replace the suffocating girl in the *Dirty Harry* movie with a nuclear bomb on Manhattan, and replace Scorpio with a captured member of a terrorist team, and you get the version of the ticking bomb scenario that regularly informs the plot of the TV series *24*. Its protagonist, agent Jack Bauer of the Counter Terrorist Unit (CTU) in Los Angeles, is regularly involved in torture scenes to ward off atrocities that would otherwise be imposed by terrorists with weapons of mass destruction. Torture is invariably presented as unfailingly effective, except, of course, when it is occasionally inflicted on Bauer himself (Mayer 2007). The moral urgency of time running out is dramaturgically reinforced by a reappearing ticking digital clock on the screen. The impact is aptly described by Jane Mayer as 'a riveting sensation of narrative velocity'.

Alas, catastrophic moral horror is also the starting point for Alan Dershowitz (2002), the well-known Harvard law professor and criminal defence lawyer, who has frequently invoked the scenario in American debates on the justifiability of torture or torture-like interrogation methods in the War against Terror.

It is important to appreciate the fact that what gives Dershowitz's example a certain intuitive appeal is *not* that it is a traditional utilitarian justification in terms of balancing costs and benefits; rather, it is a *right-based* justification, which recognises that *information* obtained by torture can never be presented as *evidence* in court. However, we should resist the application of this thought-experiment – because *that* is precisely what it is: an ethical *thought-experiment* – to real-life situations. Although the right of suspects not to be harmed cannot be more stringent than the right of innocent bystanders not to be killed, these rights collide in so extreme (and extremely rare) cases that no *legislator* can be justified in legally authorising the numerous 'Jack Bauers' to inflict pain and injury on the alleged 'unlawful combatants' in the circumstances characterising the fight against terrorism. It does not follow, from the moral intuitions underlying the 'ticking bomb scenario', that officials, in real-life situations, are entitled to ignore moral side-constraints in their pursuit of noble ends.

A problematic aspect of Dershowitz's 'ticking bomb scenario' is that it is used, in public debates, to justify something else, namely the ongoing detention and interrogation practices at Guantanamo Bay, and the outsourcing of torture and torture-like techniques, inflicted on an unknown number of 'ghost detainees' indefinitely detained at an unknown number of interrogation centres elsewhere in the world. The current US administration has responded to widespread national and international criticism by claiming that unlawful combatants are, in the words of former Attorney General John Ashcroft, 'not entitled to and do not deserve the protections of the American constitution'.

Basically, this is yet another illustration of how the logic of pre-emption has been established as a new paradigm for dealing with uncertainty, as something distinct from quantifiable risks. Nevertheless, the paradigm of pre-emption can be compatible with the rule of law only to the extent that it incorporates the principle of *habeas corpus*, which is the fundamental right of citizens and aliens to challenge, in independent courts, the factual and legal basis for detention, imprisonment and other coercive practices by the state. Here, the distinction between information and evidence can no longer be maintained. A democratic state that fails to respect the principle of *habeas corpus* is a state that has ceased to be a constitutional democracy.

THE RELEVANCE OF RIGHTS

At least prior to September 11, rights occupied a central place in public discourse on law and morality. In a comprehensive survey of police research, *Policing Citizens* (1999), P.A.J. Waddington plausibly demonstrates the *explanatory* significance of human rights for policing by way of contrasting the British colonial police's paramilitary subjection of native populations in the former colonies with the relatively constrained exercise of police authority in domestic affairs. Unlike people in the colonies, people in the mainland were recognised by the police as citizens with rights, indicating that

> . . . *how* a society is policed depends upon *who* is policed. When it is citizens with civil and political rights, then policing is approached with caution; but when the recipients of police authority are not citizens, then police are free to exercise naked coercive force.
>
> (Waddington 1999: 26, italics in original)

Of course, in a *justificatory* context, state officials are hardly 'free to exercise naked coercive force'. Rather, the moral and legal challenge of human rights flows from the fact that non-citizens no less than citizens are bearers of certain fundamental rights. These rights entitle their bearers to basic respect and recognition, committing state officials to provide justifications whenever their exercise of coercive authority interferes with the autonomy of persons. Accordingly, the moral and legal significance of Waddington's explanatory dichotomy between citizens and non-citizens extends to the domestic domain as well, because, as he aptly points out:

> . . . policing is not simply restrained or unrestrained *per se*, but tends to be restrained when dealing with *some* members of the civil population and less so when dealing with others . . . If citizenship is unevenly

distributed, the civil population are not merely passive recipients of rights: they assert their citizenship and contest its denial.

(Waddington 1999: 28–29, italics in original)

Thus, within a state's jurisdiction, human rights constitute moral and legal yardsticks enabling citizens and non-citizens to identify and critically assess contested exercises of state authority. Critical assessment of state practices in terms of human rights involves three separate standards. First, a state, by ratifying an international human rights convention, commits its officials to passively *refrain* from violating the rights of persons (and, infrequently, the rights of collective entities) comprised by the convention. Second, officials of a ratifying state are committed to actively *protect* those among the convention's recipients who are unable to defend their rights against illegitimate interference by third parties.[2] The combined obligation to 'respect and protect' does not undermine the well-known distinction between negative and positive rights. As Alan Gewirth observes, a police officer's duty to protect (and not merely respect), say, the negative right not to be murdered is a duty 'to see to it that potential offenders *refrain from* the prohibited actions' (Gewirth 1996: 35, author's italics). Consequently, although the content of a correlative duty to defend a negative right necessitates a positive act of interference, the right itself is nevertheless properly classified as negative. Third, positive rights correlate with duties to perform two separate positive actions: not merely a duty to protect the right-holder's enjoyment of the good or benefit against illegitimate attacks from third parties but also a duty to make the good or benefit available to the right-holder in the first place.

The contents of rights are goods or benefits to which right-holders attach great moral significance; the right, for example, to life, liberty and security of person, which are goods protected by the European Convention on Human Rights. Thus, by virtue of their contents, human rights are moral rights, but simultaneously they are also internationally binding legal rights, by virtue of the way they have been enacted within the UN and European systems.[3] Although the holders of human rights are almost invariably individual

2 The latter aspect of a state's dual obligation 'to respect and to ensure' human rights resonates strongly with moral sentiments of police culture. The solidarity felt by many police officers with victims of crime is cogently expressed by the recurrent remark of Detective Frank Pembleton (powerfully played by actor Andre Braugher), the ruthless interrogator in the acclaimed NBC television series *Homicide: Life on the Street:* 'We speak for those who cannot speak for themselves.'

3 Among human rights scholars, there are some disagreements about the extent to which the Universal Declaration has attained the force of law during the more than 50 years that have elapsed since its adoption in 1948 by the UN General Assembly. This issue, however, has no bearing on the status of moral/human rights regulating the use of force and violence by state officials, which are extractable from covenants adopted by the UN and the Council of Europe, respectively.

human beings,[4] the addressees of those rights are the ratifying states, which undertake, as already mentioned, a three-fold obligation to 'respect, protect and provide' the various goods picked out by the corresponding rights.

Infringements of rights

Theoretically, almost *any* right included in human rights covenants might necessitate the use of force and violence by the police and other agencies of the state in order to protect right-holders against illegitimate interference. Moreover, even the provision (and not merely the protection) of a positive human right occasionally implies the use of physical force and violence against some among its recipients, undermining their active enjoyment of benefits guaranteed by negative as well as positive rights. Consider, for example, the *possible* conflict between article 12 of the International Covenant on Economic, Social and Cultural Rights, which stipulates the positive 'right of everyone to the enjoyment of the highest attainable standard of physical and mental health', and article 18 of the International Covenant on Civil and Political Rights, which states that 'Everyone shall have the right to freedom of thought, conscience and religion'. In health systems of western welfare states, the former right more or less routinely requires, on certain medical indications, the administration of blood transfusions on consenting patients and patients whose physical or mental conditions require consent by their legal representatives. Yet there are US court cases in which members of Jehovah's Witnesses, who oppose blood transfusion on religious grounds, have failed to substantiate their alleged status as victims of illegitimate state coercion. For example, in *Raliegh Fitkin-Paul Morgan Memorial Hospital v Anderson* (1964) the court ordered a blood transfusion to save the life of a woman and her unborn child. It is easy to imagine a quite plausible scenario in which the enforcers of the court's decision must resort to the use of physical force and possibly even violence to subdue a vigorously resisting patient, trespassing her negative right to manifest her 'religion or belief in worship, observance, practice and teaching'. However, the court's justification for discounting the patient's right to freedom of religion is recognised by paragraph 3 of article 18:[5]

> Freedom to manifest one's religion or beliefs may be subject to such limitations as are prescribed by law and are necessary to protect public

4 An exception is the right of self-determination, which applies to peoples, not persons.

5 Needless to say, paragraph 3 of article 18 was not available to courts in 1964, two years prior to the covenant's adoption (in 1966) by the General Assembly. In fact, the Civil Rights Covenant was not ratified by the United States until 1992. Because the USA has to date not adhered to the covenant's Optional Protocol, it is not possible for victims of human rights violations to bring complaints against the USA before the Human Rights Committee established by the Civil Rights Covenant.

safety, order, health, or morals or the fundamental rights and freedoms of others.

In other words, the court's order to coercively (and possibly forcefully or violently) ensure the patient's positive right to physical health was, in retrospect, justifiable in terms of the unborn child's right to life,[6] despite the fact that discharging two human right duties (to protect life and provide health) was tantamount to discarding a third (to respect religious freedom). Importantly, in addition to rights, paragraph 3 refers to vital goods like public safety, order, health and morals, only two of which (health, order) are incorporated as rights in UN covenants. This juxtaposition of goods as distinct from rights is plausibly taken to imply that whilst the former are social *goals* to be promoted, the latter are constraints that may or may not be overridden by either goals or other, more stringent rights.

Legitimate aims and conditions of infringement

In the terminology of the European Court of Human Rights, the pursuit of goals as well as the protection of (more stringent) rights are *legitimate aims* which may or may not justify a state's imposition of limitations on right-holders' exercise of their rights. Consequently, from the point of view of political philosophy, the term 'legitimate aims' comprises 'the good' no less than 'the just'.[7] Article 18 belongs to a group of human rights that not only picks out particular goods for protection but subsequently proceeds to explicitly enumerate the conditions under which right-holders are justifiably deprived of these goods. Other examples of what might be called internal derogation are article 2 (life), 5 (liberty), 8 (privacy), 9 (freedom of thought),

6 True, paragraph 3 refers to the rights of *others*, not the rights of the individual whose rights are infringed. It might be the case that respect for the patient's way of practising her religious beliefs constitutes a constraint on what the state is allowed to do to save her life. To refrain from providing the necessary blood transfusion is in some sense tantamount to 'performing' (by omission) a kind of passive euthanasia. According to the doctrine of doing and allowing there is a morally relevant difference between bringing about some harm and letting some harm befall someone, which might explain why passive and not active euthanasia is morally permissible. On this interpretation of article 18, the best justification for the court's decision is the unborn child's and not the mother's right to life.

7 In *Ethics for Adversaries* (1999), Arthur Applbaum has 'at the risk of invoking the title of a spaghetti western . . . identified three orders of reason: the good, the just and the legitimate' (p. 217). Citizens of pluralistic societies frequently fail to reach agreement in their deliberations about the good, and such disagreements must be resolved at the level of justice. Alas, there are rivalling conceptions of what justice requires as well. For example, citizens disagree as to whether women should be granted a right to abortion. In the light of possibly irreconcilable disagreement as to what justice requires, citizens may agree to defer to the legitimate authority of Parliament. Evidently, as Applbaum observes (p. 218), 'at some point one runs out of reasons'.

10 (freedom of expression) and 11 (freedom of assembly) of the European Convention on Human Rights. Thus, when some of these rights, like the right to life, are said to be non-derogable, the reference is to the *external* derogation applying to the majority of rights in virtue of article 4 of the International Covenant on Civil and Political Rights and article 15 of the European convention 'in time of war or other public emergency threatening the life of the nation'. However, with the exception of the prohibition against torture, no human right is absolute, the conditions of derogation being either internally and externally explicated by rights or extractable from covenants by interpretative reconstruction of the scope and stringency of rights.

Although conditions of internal, external and extractable derogation are conditions of the legitimate infringement of rights, the UN covenant nowhere explicitly endorses infringement by the use of force and violence. The European convention does so only once, in article 2 on the right to life, which in paragraph 2 states:

> Deprivation of life shall not be regarded as inflicted in contravention of this article when it results from *the use of force* which is no more than absolutely necessary: a. in defence of any person from unlawful violence; b. in order to effect a lawful arrest or to prevent the escape of a person lawfully detained; c. in action lawfully taken for the purpose of quelling a riot or insurrection (italics added).

To be sure, it is theoretically feasible to have a system of law devoid of physical enforcement mechanisms. One of the fundamental functions of rules is, as Wittgenstein pointed out in his celebrated analysis of what it means to follow a rule in *Philosophical Investigations* (1953), to establish standards enabling practitioners (and observers) to distinguish between a right and a wrong way of doing things. Thus, to follow a rule logically implies the possibility of making mistakes. A purely declarative system of legal rules would suffice to fulfil this function, providing its subjects with authoritative criteria of appraisal and criticism. Indeed, the UN and European systems of human rights are declarative in the sense that decisions by, respectively the UN Human Rights Committee and the European Court of Human Rights in cases of individual complaints are entirely dependent on domestic enforcement by ratifying Member States.[8]

8 At the international level, the Security Council and the General Assembly, which are UN charter-based institutions, have repeatedly been criticised for declarative impotence, failing to fulfil their obligation 'to take effective collective measures for . . . the suppression of acts of aggression or other breaches of the peace' (UN charter, article 1, paragraph 1). Most significantly, article 2(4) prohibits 'the threat or use of force' by Member States 'against the territorial integrity or political independence of any state'. Nevertheless, under article 51 each Member State retains its 'inherent right of individual or collective self-defence' against

However, when states explicitly take on an obligation to 'respect and to ensure' and to 'secure to everyone within their jurisdiction' the rights included in the UN and European conventions, domestic enforcement would be radically self-defeating and subversive of these aspirations to the extent that use of force and violence were to be excluded in the absence of positive textual endorsement. Consequently, it seems plausible to argue that conditions of internal, external and extractable derogation of human rights comprise the conditions of the legitimate infringement of these rights by the state's use of force and violence.

TWO LEGAL PARADIGMS

In order to reinforce the conclusion that 'the ticking bomb scenario' does not constitute such conditions of legitimate infringements of the prohibition against torture in the fight against terrorism, it might be fruitful to situate the events of September 11 in a more comprehensive conceptual framework. Two competing *legal* paradigms are frequently invoked by state officials and commentators. According to the first paradigm, the atrocities were the effects of an 'armed attack' on America, justifying President Bush's subsequent declaration of 'war against terrorism'. According to the second paradigm, the atrocities constituted 'crimes against humanity', implying that a law enforcement response would be more appropriate. Intuitively, the phenomenon of war involves large-scale, institutionalised forms of force and violence which conflicting political entities inflict on each other. Clearly, John Keegan is right to point out in *A History of Warfare* (1993: 3) that the practice of warfare 'antedates the state, diplomacy and strategy by many millennia'. Yet in its modern version, war and warfare are intimately connected to the international system of nation states. This system is, despite political realism's allegations to the contrary, essentially a normative and moralised system, constituted by the mutual recognition (and sometimes rejection) of normative claims articulated by officials of states. To a large extent, such claims are either recognised as principles of customary international law or increasingly enacted as positive international law, stipulating standards of right and wrong conduct. This implies that we are not, as observers, completely free to

military aggression by adversarial states. Moreover, under article 42 the Security Council may authorise 'such action by air, sea, or land forces as may be necessary to maintain or restore international peace and security'. However, it is a contested issue whether such measures, which evidently must encompass the collective use of force and violence against a state's territorial integrity, can be authorised by the Security Council to prevent large-scale human rights violations by state governments against segments of their own populations. In contrast, *peacekeeping* involves, with the consent of a target state, the use of UN military and police personnel to keep rivalling segments among the state's population apart. In peacekeeping operations, UN officers are permitted to use force and violence in self-defence only.

identify phenomena like war and warfare as we please. Rather, because war and warfare have become thoroughly legalised and, by implication, also thoroughly moralised, we can do no better than to attempt to reconstruct the framework of international law and see whether it can appropriately accommodate the events of September 11.

As Michael Walzer (1977: 21) observes, war is morally judged twice, first with respect to the reasons states have for going to war in the first place (*jus ad bellum*: the justice of war) and second with respect to the way the war is waged or what kind of methods are used in the waging of war (*jus in bello*: justice in war). Of course, *jus ad bellum*-considerations are crucial ingredients in the just war tradition. A basic distinction is made between war as *self-defence* against unjust aggression and war as the infliction of *punishment* of state wrongdoing. The latter category is deeply problematic and relies on a flawed analogy with the guilt of individual wrongdoers in a way that the conception of war as self-defence does not.

The impact of just war thinking on the normative regulation of international conduct is indicated by the fact that just war principles have to a large extent been codified as positive international law. Yet the principle that war is justified as the infliction of punishment of wrongdoing is not one of them. On the contrary, the United Nations' Security Council has repeatedly made crystal-clear that retributive justice by way of *reprisals* is absolutely at odds with the basic principles of the UN charter. Therefore, the argumentative strategy adopted by the USA and Israel, the two foremost practitioners of reprisals on the international scene, has been, as Christine Gray (2000: 119) has observed, to stretch the meaning of self-defence so as to include the right to inflict reprisals. So far, these attempts to re-describe the infliction of reprisals as justifiable self-defence have failed to win the explicit approval by other members of the United Nations.

On September 12, the NATO allies stated that the events of the preceding day constituted an attack on *all* the members of the alliance in terms of the mutual defence guarantee of the treaty's article 5. Of course, one might argue that the NATO members in a similar way stretched the meaning of collective self-defence in article 5. To be sure, when article 5 was drafted 52 years ago, no one could, in the words of Philip Gordon (2001: 89), 'have imagined that its first invocation would involve Europeans coming to the aid of the United States rather than the other way around'. As many commentators have pointed out, a statement by the NATO Council is, taken in isolation, hardly sufficient to establish the conclusion that the United States was entitled to invoke the right of individual self-defence within the meaning of article 51 of the UN Charter. This right is said to be *inherent* in the sense that it can be exercised by members of the United Nations independently of explicit authorisation by the Security Council, provided that 'an armed attack' has occurred. All *other* exceptions to the general prohibition of the threat or use of force contained in article 2(4) of the UN Charter must be authorised by the Security Council.

However, to the extent that it could be legally established that the events of September 11 constituted 'an armed attack', it seems reasonably clear that international law did provide the United States and its allies with a just cause for waging war in Afghanistan.[9]

Warfare or law enforcement?

Nine days after the attacks on the World Trade Center and the Pentagon, President Bush made a sweeping statement in his speech to Congress: 'Our war on terror,' he said, 'begins with al-Qaeda, but it does not end there. It will not end until every terrorist group of global reach has been found, stopped and defeated' (Stevenson 2001: 35). However, the limited applicability of the paradigm of warfare in the fight against terrorism was recognised even by then Secretary of Defence Donald Rumsfeld, who indicated that the military elimination of every terrorist would be 'setting a threshold that is too high'. Indeed, the strictness of conditions under which a war against terrorism can be justifiably waged serve to illustrate the primacy of the second legal paradigm, the paradigm of law enforcement and criminal justice.

Nevertheless, although warfare and law enforcement are plausibly considered as distinct ideal types in Weber's sense, they sometimes tend to converge metaphorically in terms of slogans like 'war against crime' or 'war against drugs'. Because such metaphors resonate strongly with certain characteristics of police culture, they have real consequences, as demonstrated by Jerome Skolnick in his classic study *Justice without Trial* (1966). The basis for Skolnick's illuminating reflections on police attitudes towards law and legal reasoning is Herbert Packer's (1968) influential discussion of two models of the criminal justice system, which he called the 'due process model' and the 'crime control model'. The due process model embodies the doctrine of the rule of law, emphasising individual rights and legal restraints on the various governmental agencies operating within the criminal justice system. The crime control model, meanwhile, aims at maximising the criminal justice system's ability to discover arrest, prosecute, convict and incapacitate criminal wrongdoers, and hinges on the assumption that formal law enforcement is the most efficient way of controlling or reducing the overall level of crime. The challenge, as Packer saw it, is to strike a proper balance between those two models; to achieve 'order through law'.[10]

9 But, needless to say, *not* in Iraq.

10 Yet, as Andrew Ashworth (1994: 28–29) has pointed out, Packer's distinction is too rigid in certain respects. Ashworth lists five objections, to which I would like to add the following two: (1) notions of fairness, so crucial in most accounts of due process, are embedded in criminal law in terms of *mens rea* requirements for criminal liability; (2) violations of due process norms sometimes generate criminal liability and thus should be counted among the crimes to be controlled or reduced according to the crime control model.

However, a fundamental feature of police culture, originally conceptualised by Skolnick and subsequently confirmed by anthropologists like Jonathan Rubinstein (1973) and Malcolm Young (1991) among others, is likely to imply that the scales of criminal justice are already tilted in favour of the model of crime control. A crucial passage in Skolnick's study explains why this is so:

> . . . it must be understood that the police draw a moral distinction between criminal law and criminal procedure. . . . The substantive law of crimes is intended to control the behavior of people who wilfully injure persons or property, or who engage in behaviors eventually having such a consequence, as the use of narcotics. Criminal procedure, by contrast, is intended to control authorities, not criminals. As such, it does not fall into the same moral class of constraints as substantive criminal law.
>
> (Skolnick 1966: 191–192)

It is important to realise that this distinction of 'police cosmology' between criminal law and criminal procedure is, as Skolnick points out, a morally based distinction, despite the fact that it frequently emerges as cynicism or disillusionment about the realities of the criminal justice system. Above all, police cynicism reflects the view that procedural norms are unrealistic and counterproductive external constraints to be stretched or set aside in the higher mission of fighting crime, maintaining order and protecting the rights of victims.

Cynicism about procedural rules, then, can hardly be understood independently of the strong moral commitments of crime fighting. A typical expression of this sense of moral mission is found in the preface to the American *Law Enforcement Handbook*:

> As police officers, we have a vital role: we are the front-line troops in the war against the forces that would disrupt society and destroy the right of our fellow citizens to live in peace and security. Ours is an honourable task, one that requires our best.
>
> (Rowland and Bailey 1985: 13)

Not surprisingly, then, whether or not moral conditions for waging a just war against terrorism can be met, actual warfare will have a profound impact on legislation and the theory and practice of law enforcement. Dramatic changes in the legal framework of criminal justice agencies have been introduced in several countries, most notably in the United States itself, in the form of the so-called PATRIOT Act, and in the United Kingdom, in the form of the Anti-Terrorism Crime and Security Bill. Enacting repressive modifications in the legal frameworks of constitutional democracies is not necessarily at odds with fundamental principles of human rights and the rule of law. As already mentioned, article 15 of the European Convention on Human Rights allows derogation from certain rights, including the right to liberty and security of person (article 5), 'in time of war or other public emergency threatening the life of the nation'. The right to liberty and security of person was precisely the right that the British government derogated from when its Security Bill, with some minor modifications, was passed. Moreover, according to the influential human rights organisation JUSTICE (November 2001: 4), 'a court would be likely to find, in the current circumstances, that a public emergency does exist for the duration of any credible threat from the Al Qaida organisation'.

FIGHTING TERRORISM: SECURITY VERSUS LIBERTY?

It is quite common to think of such derogation as the result of a necessary trade-off between, on the one hand, the vital interests of security and, on the other hand, the no less vital interests of liberty. Yet this way of thinking about security and liberty tends to ignore the urgent need for legal safeguards whenever liberty is set aside for the sake of security. A more principled basis for thinking about these matters is suggested by John Rawls' requirement that arguments for coercive restrictions of liberty must proceed from the principle of liberty itself. 'Limitations upon the extent of liberty,' Rawls says in *A Theory of Justice*, 'are for the sake of liberty itself and result in a lesser but still equal freedom' (Rawls 1971: 247). According to Rawls' moral individualism, which insists that the moral status of the state is parasitic upon the moral status of persons and not the other way around, a state can justifiably invoke the principle of national security only to the extent that its application of that principle is essentially informed by the individual rights and values that are to be secured.

It is very hard to see that this is the case with the establishment of military tribunals, authorised to secretly try and possibly sentence to death non-citizen terrorist suspects, suspending standard requirements of evidence and the right of appeal. The relevant answer to Ashcroft's statement that foreign terrorists 'are not entitled to and do not deserve the protections of the American constitution' is, of course, that the argument is circular, presupposing that the guilt of suspects has already been established independently of fair procedures in general and defensible interrogation techniques in particular. It is also very hard to see that the principle of 'restricting liberty for the sake of liberty' is properly reflected in various provisions of European legislations, allowing for more or less indeterminate detention of foreign suspects without trial. Such measures become particularly worrisome when considered in the light of Thomas Mathiesen's convincing documentation (2002) of a consistent drift among members of the European Union towards expanding the very concept of terrorism.

Significantly, there is a realistic alternative to all of this, as indicated by the fact that American federal courts, without compromising fundamental principles of constitutional democracy, have tried and convicted persons with links to bin Laden and al-Qaeda for the 1993 bombing of the World Trade Center (killing six people) and the 1998 bombings of the American embassies in Kenya and Tanzania (killing 224 people). As these trials illustrated, there are, in terms of domestic criminal law, a vast number of offence categories that might apply to the events of September 11.

However, there is one category of customary international law that seems to be particularly relevant, namely the concept of 'crimes against humanity', which appeared for the first time in 1945, at least as a technical legal term, in the charter of the Nuremberg trial. 'Crimes against humanity', as well as war crimes, have been at the forefront at the International Criminal Tribunals for the former Yugoslavia and Rwanda, established under the authority of the Security Council. Yet according to Law Professor Ruth Wedgwood (2001), 'only 31 individuals have been tried by the Yugoslav tribunal in 8 years, at a cost of $400 million', so it seems to be a more feasible alternative to let national courts exercise their so-called universal jurisdiction over crimes such as genocide, crimes against humanity and war crimes.

CONCLUSION

Let me, in conclusion, very briefly situate my discussion of torture, terror and rights within the grand historical context of Samuel Huntington's theory of 'clash between civilisations', which has frequently been invoked by commentators in the wake of September 11. I take this theory to be empirically false, but I shall not justify that claim here. However, in contrast to theories of natural science, the objects of the social sciences and the humanities are, as Charles Taylor (1985: 45) puts it, 'self-interpreting animals', capable of making false theories true by acting on them. There are some significant and destructively influential agents on the international scene who do subscribe to the theory of 'clash between civilisations', namely bin Laden and whatever is left of his al-Qaeda organisation and the Taliban regime. Yet it is not merely up to them whether the clash will become true. It is also up to the western countries to exercise moral constraint, when waging war and pursuing criminal justice, to ensure that the theory is performatively falsified.

References

Applbaum, A.I. (1999) *Ethics for Adversaries*, Princeton: Princeton University Press.
Ashworth, A. (1994) *The Criminal Process*, Oxford: Oxford University Press.
Caro, R.A. (1974) *The Power Broker*, New York: Knopf.

Dershowitz, A. (2002) *Why Terrorism Works*, New Haven and London: Yale University Press.
Evans, M. and Morgan, R. (1998) *Preventing Torture*, Oxford: Clarendon Press.
Gewirth, A. (1996) *The Community of Rights*, Chicago: University of Chicago Press.
Gordon, P. (2001) 'NATO after 11 September', *Survival*, 43(4).
Gray, C. (2000) *International Law and the Use of Force*, Oxford: Oxford University Press.
JUSTICE (2001) 'Briefing on the Anti-Terrorism Crime and Security Bill', London.
Keegan, J. (1993) *A History of Warfare*, New York: Vintage Books.
Klockars, C.B. (1985) 'The Dirty Harry Problem', in F. Elliston and M. Feldberg (eds) *Moral Issues in Police Work*, Lanham: Rowman & Littlefield Publishers.
Mathiesen, T. (2002) 'Expanding the concept of terrorism?', in P. Scraton (ed) *Beyond September 11: An anthology of dissent*, London: Pluto Press, 84–93.
Mayer, J. (2007) 'Whatever it takes', *The New Yorker*, 19 February.
McCoy, A. (2006) *A Question of Torture*, New York: Metropolitan Books.
Nozick, R. (1974) *Anarchy, State, and Utopia*, Oxford: Basil Blackwell.
Packer, H. (1968) *The Limits of the Criminal Sanction*, Stanford: Stanford University Press.
Rawls, J. (1971) *A Theory of Justice*, Oxford: Oxford University Press.
Rowland, D. and Bailey, J. (1985) *The Law Enforcement Handbook*, New York: Barnes & Noble Books.
Rubinstein, J. (1973) *City Police*, New York: Farrar, Straus & Giroux.
Skolnick, J. (1966/1994) *Justice without Trial*, New York: MacMillan College Publishing Company.
Stevenson, J. (2001) 'Pragmatic Counter-Terrorism', *Survival*, 43(4).
Taylor, C. (1985) 'Self-interpreting animals', in C. Taylor, *Human Agency and Language*, Cambridge: Cambridge University Press.
Waddington, P.A.J. (1999) *Policing Citizens*, London: UCL Press.
Walzer, M. (1977) *Just and Unjust Wars*, New York: Basic Books.
Wedgwood, R. (2001) 'Tribunals and the events of Septrember 11th', in *ASIL Insights*, Washington: The American Society of International Law.
Wittgenstein, L. (1953) *Philosophical Investigations*, Oxford: Blackwell.
Young, M. (1991) *An Inside Job*, Oxford: Clarendon Press.

Epilogue

The inescapable insecurity of security technologies?[1]

Lucia Zedner

INTRODUCTION

Techno-credulity – or blind faith in technological solutions to otherwise irresolvable problems – is a hallmark of late modernity. Enormous trust is placed in the capacity of technology to surmount the gravest challenges to our well-being and happiness. The big hazards of disease, poverty, global warming, and, not least, insecurity are the foci of intense technological innovation dedicated to curing the incurable, eradicating the ineradicable, securing the vulnerable, and, of course, saving the planet. The political, financial and emotional investment in the cure-all capabilities of modern technology is on such a grand scale as seriously to inhibit critical scrutiny of the tensions, ironies and paradoxes that arise in seeking technological solutions to the ills of modern life (Marx 2001). This epilogue reflects and builds upon the superb contributions to this volume that together institute just such a critical enquiry into the insecurities of technologies whose very raison d'être is ostensibly to provide security.

The preceding chapters make clear the extraordinary variety of technologies capable of being viewed through the lens of security. The technologies discussed extend from the mundane to the exceptional: from the 'everyday objects' discussed by Neyland to the torture-like interrogation techniques explored by Halvorsen. They range from high- to relatively low-tech. Contrast the satellite tracking of offenders (Nellis) to the national ID cards debated by Lyon. They comprise virtual technologies, for example access to the internet (Jewkes) and computer crime (Yar), and concrete ones like the physical spaces and architecture considered by Jones. And their deployment ranges from the expert to the amateur. Compare the operators of CCTV systems (discussed by Smith and by Goold) and the use of DNA (Dahl) to the amateur photographers whose surveillance activities are considered by Koskela. Clearly the range of technologies to which security gives its imprimatur is far-reaching.

1 I am grateful to Abigail Bright for her excellent research assistance, and to Ben Goold and the editors for their insightful comments.

And yet their common pursuit of security gives a greater conceptual coherence to this book than the diversity of tools and techniques under discussion might at first suggest.

Especially welcome is the fine-grained analysis of particular substantive examples that each contribution pursues. Too often the literature on security and surveillance remains at the macro-theoretical level. Relying upon its distance from the details of everyday usage it ignores fine distinctions and significant divergences, and so licenses sweeping generalisation. Ideal types and newly minted abstractions abound: witness, 'the new surveillance (Marx 2004), the 'surveillant assemblage' (Haggerty and Ericson 2000) and the 'surveillance web' (McCahill 2002). These inventive coinages attempt to capture the scale and magnitude of change wrought by the rise of surveillance technologies and their impact upon social and political life. But in seeking to grasp the enormity of the transformation, they often fail to pay sufficient attention to the particular and the problematic. What is most welcome about the contributions to this volume is their melding of theoretical insight with close attention to the concrete and the local. In their irreducible diversity the technologies under discussion, their varied institutional and operational settings, differing goals, diverse cultural milieus and divergent meanings reveal the full complexity of surveillance. Most importantly, the chapters make clear the limits of technology and its inherently problematic role in seeking the elusive goal of security.

As the subtitle of this book makes evident, although the technologies in question are principally driven by insecurities, they are also products of, and contribute to, the larger development of surveillance. Reference to 'everyday life' is particularly apt. Where once surveillance was the stuff of covert undercover operations by state security and intelligence services and its subjects were the spies and spooks of the Cold War era (Marx 1989), increasingly surveillance has been normalised, it is ubiquitous, and its gaze extends to all. Surveillance is an embedded, commonplace facet of everyday life in ways that were unimaginable 50 years ago and its technologies are dispersed through our social world. Surveillance technologies that were once the province of experts within elite state institutions, like intelligence bureaux and counter-espionage agencies, have been democratised and are now readily available for purchase by ordinary personal consumers.[2]

The result is that our everyday physical environment has become peppered with the tools and techniques of surveillance, transforming urban architecture and the way we move through it. In radically different ways the chapters in this book reveal the strongly spatial character of contemporary surveillance. The embedded quality of security reveals itself particularly in locales of high

2 The extent to which they have percolated down is nicely illustrated by the fact that for her last birthday, our seven-year-old daughter was bought a working remote surveillance camera construction kit by a classmate.

risk such as airports (Neyland, Jones) and the spatial logic of security operations becomes most clearly manifest in the organisation of major events (Klauser). Well-established architectural techniques become an important armoury in the erection of the boundaries, barriers, gateways, checkpoints and multiple other structural means by which to manage otherwise unsecured flows of people.

The rate of innovation and rapid spread of surveillance technologies has had such transformative effects upon communication and transport, social, political and economic life that it is easy to fall prey to technological determinism. Social problems and their resolution appear to reside in technology alone. The socially situated nature of technology, the political and moral choices it entails and the economic costs and consequences of these choices are only now being scrutinised. The manner in which technologies are applied, the 'relationships of power, gender, race and global economic inequality' that permeate access to technologies (Aas 2007: 154) and the resultant inequalities of application and take-up are ripe for further enquiry. The sociological tendency to see society as shaped by technology (Harcourt 2007: 33) needs to be countered by critical scrutiny of the profoundly political character of technological innovation and deployment. Even if we accept that technology itself is neutral (a contestable premise in itself), it is shot through with social meaning, it is politicised and filtered through the cultural lens of those applying it in ways that technological determinist accounts belie (Goold, 2004: Chapter 8).

Techno-credulity is widespread. In a recent book Harcourt ascribes the present dominance of actuarial technique in criminal justice policy to determinism, arguing 'it chose us' (Harcourt 2007: 32). But, as I have argued elsewhere: 'to ascribe major shifts in penal policy solely to technological possibility may be to place the proverbial cart before the horse. The causal relationship between the technology of profiling and predicting and the political demand for public protection is surely more complex than the simple assertion that actuarialism "chose us" ' (Zedner 2008). Claims of technological takeover such as these rely upon a curious denial of human agency, attention to which requires us to consider other causal factors that may render it altogether less powerful than at first appears. The terrifying technological possibility of total surveillance, promised for example by CCTV, may be undermined by the less exciting actuality that no one is watching or, at least, that no one is watching all the time (Lyon 1994: Chapter 1; Goold 2004: Chapter 1).

The contributors to the present volume, all leading experts in their fields, have done much to redress the tendency to technological determinism by exercising their collective critical faculties upon the claims made for technology in respect of security broadly defined. This short epilogue cannot hope to do justice to the richness and diversity of these contributions. Instead it seeks to question the claims made for new technologies and in so doing to make clear the limits of security on offer. It seeks also to elicit from the preceding chapters common themes not immediately apparent on first reading and to expand upon their implications for social life. Finally, it seeks to tease out conflicts and contradictions, inconsistencies and discrepancies in the ways technologies of security operate and are employed.

THE LIMITS TO TECHNO-CAPACITY

In the introduction to this volume, the editors embrace an extended notion of technology, noting 'the term technology in this context obviously applies to a broad social matrix of action which enables humans to modulate their environments' (Introduction). Quite how *enabling* technology is, despite its implicit, and often explicit, promise to empower its users to alter and adapt their social world is a moot point. Part of the problem is that precisely because technology extends beyond the relatively predictable capacities of physical gadgets and electronic wizardry to entail its deployment by human operators, in social settings, and over or against human subjects, basic assumptions about its capacities commonly prove to be ill founded. Technologies are mediated through professional sub-cultures, as well as through a more amorphous 'safety culture' that differs radically in different social settings.[3] They are also mediated through the highly complex dynamic entities that are organisations. The attempt to think holistically about the interplay between technologies, professions and organisations generates new territories of enquiry. The emergence of so-called 'safety management systems' has been a particularly important development in this respect and has become a major topic of research (as witnessed by journals such as *Safety Science*. See also Hale and de Kroes 1997). This trend has been further fuelled by increasing emphasis in European directives on auditable safety management systems and the development of the precautionary principle as a framework for decision making in the absence of scientific knowledge (Fisher 2002). By way of illustration let us examine the deployment of safety management systems in loci of high risk like air safety. This example is deliberately chosen as a hard case by which to test the claims made for technology and through which to explore the limits of even the most sophisticated technological endeavour to deliver its promise of security (see also Jones, this volume).

But first some reflections on the origins of safety management systems. They were developed originally in fields like engineering, where it is arguably more plausible to attempt to calibrate, and by so doing to manage, the risks that might otherwise arise as a consequence of failure within complex

3 The term 'safety culture' is ubiquitous in environmental and engineering literatures, though used in very different ways in each. It is said first to have appeared in the report on the 1986 Chernobyl disaster.

technological systems (such as nuclear reactors). In order to avert, or at least to minimise the risk of, systems failure it is necessary to have comprehensive knowledge of the entirety of that system. Upon the basis of this knowledge probabilistic assessment can be made of the likelihood of various kinds of systems failure.

In scientific usage, the term system has the well-defined meaning of a set of relations so connected that deductions can be made from one set of relations to others or from relations among the various entities to the behaviour of the system as a whole. Scientists also use the term system to denote an entity that is explicitly distinguished from its environment, whose internal elements are clearly defined, and whose internal and external relationships can be unambiguously stated (Feeney 1985: 8). The problem with this scientific definition is that it sets up criteria so demanding it is doubtful that they could be fulfilled by the open-ended interactions of human agency that characterise even supposedly 'hard science' environments. Where the system is open-ended, such as is the case in respect of toxicology or epidemiology, the efficacy of probabilistic assessment is reduced by the difficulty of determining what risks are in play. In respect of air transport security, it is questionable whether it makes sense to talk of a 'system' at all.

The very notion of a safety management system thus reflects the distance between natural and social science. Natural science derives from a fundamentally optimistic assumption that it is possible to know the world and by knowing alter, or even control, it. Scientists of the social world enjoy no such optimism: their studies constantly force them to face up to the difficult realities of trying to manage human behaviour and interaction. When safety management systems are wrenched from their natural science and engineering origins to be reapplied in social settings, such as airports, the limits to their technological capacity quickly become clear. Even in the case of closed systems, such as pertain in the field of engineering or nuclear energy, the assumption that it is possible to identify in advance all possible forms of system failure is dubious. The case of the disaster at Three Mile Island illustrates the difficulty of identifying all possible sources of risk even within the relatively closed system of a nuclear plant (Erikson 1995: Chapter 4). Whether it is remotely plausible in respect of more diffuse, social organisations is less clear still (Zedner 2006).

In respect of air safety, the use of safety management systems has been defined in a deceptively unproblematic way as 'the systematic management of the risks associated with flight operations, related ground operations and aircraft engineering or maintenance activities to achieve high levels of safety performance'.[4] Safety management systems applied to air safety pursue commendable goals: they seek to ensure that the risks pertaining in the many

4 Civil Aviation Authority Safety Regulation Group, *Safety Management Systems for Commercial Air Transport Operations: A guide to implementation*, CAP712, www.caa.co.uk, 2.

quite different sectors relating to air safety are known, that this knowledge is shared and that the collectivity of risks is properly understood and minimised. What is more doubtful is whether the promise of efficiency and absolute rationality suggested by systems analysis can be delivered in practice in this domain. Systems management radically reduces the danger that the import of key information about a putative threat is overlooked and that the implications of a risk in one sector for another sector within the system is missed. The major difficulty is that a system is by its very nature circumscribed. Once the system has been defined, information exchange within the system may be perfect but information lying outside the system is less likely to receive attention than might otherwise be the case. If key information about the planning of a terrorist threat, for example, arises outside the prescribed system, the very closed nature of the system may preclude or at least hinder such information entering the system and being given due weight. For example, the training of the 9/11 bombers by private flight schools whose small scale, commercial purpose, and geographical remoteness from the large financial and political centres of America placed them firmly outside the remit of information exchange within the safety management systems of US aviation authorities – with catastrophic consequences.

Significantly, in the attempted transfer of safety management systems to the commercial world of air safety, the authorities do not even claim to attain the rigorous standards applied in chemical plants and nuclear reactors – and this realism is to be welcomed. Rather, in adapting systems management to the very different environment of air transport they have drawn analogously upon the pre-existing features of management within large corporations. Safety management systems are thus directly 'compared with a financial management system as a method of systematically managing a vital business function' as 'an explicit element of the corporate management responsibility'.[5] Accordingly, just as risks in financial management systems are regarded as losses to be prevented through prudent financial procedures, so risks in safety management systems are also characterised as unwelcome risks to be averted by the adoption of prudent management strategies. The UK's Civil Aviation Authority guide for commercial air operators characterises the aim of the safety management system, with surprising candour, as being to:

> ensure that there are no 'business surprises'. If there are, it can be disastrous for a small company. For the larger company, unwelcome media attention usually follows an unexpected loss. An aircraft accident is also 'an unexpected loss' and not one that any company in the civil aviation industry wishes to suffer.[6]

5 Civil Aviation Authority Safety Regulation Group *Safety Management Systems for Commercial Air Transport Operations: A guide to implementation* CAP712 www.caa.co.uk, Chapter 1:2.

6 Ibid.

The safety management system is thus said to be 'as important to business survival as a financial management system'.[7] The characterisation of the loss of an airplane with all its passengers as a 'business surprise' speaks volumes about the centrality of financial considerations. And herein lies the rub, although the goal here is undoubtedly security, this is not only security in the conventional sense of safety but also, crucially, financial security. In the introduction to this book, the editors perceptively observe how many of the chapters reveal 'how security can be a façade hiding more complex causes and justifications' (Introduction). Certainly in respect of safety management systems, the not so covert goal of financial security and profit maximisation never lies far from the surface. As the UK's Civil Aviation Authority guide for commercial air operators concludes, safety management systems are 'the next step forward in safety enhancement as our industry grows'.[8] The growth and commercial success of the industry and the protection of assets is at much at issue, it would seem, as security per se.

TECHNOLOGIES AND THE CONSTRUCTION OF IDENTITY

If new security technologies have a more limited capacity to provide security than is commonly claimed for them, they nonetheless play an important part in contemporary social life. One of the most interesting themes to emerge from this collection is that technologies increasingly construct identity. Policy-makers may act as modern-day wizards conjuring new technological alchemy as the magical charms with which to cure the ills of the contemporary world. But as the Sorcerer's Apprentice learned the hard way, these spells may prove more powerful and less controllable than their makers anticipate – less charm one might say than hex. The temptation to seek technological solutions is rarely accompanied by sufficient anticipation of the fact that technologies may develop unpredictably or be subverted in ways that render them greater sources of insecurity than security. A striking contemporary example is the introduction by credit card companies of what in Britain is called 'chip and pin' (and elsewhere 'EMV') technology. This requires credit and debit card users to type in their PIN (personal identification number) instead of providing a signature at point of purchase. Introduced in order to increase security, its ubiquitous use in shops and bars significantly increases the risk that the number typed in by one customer will be seen by those queuing behind them. Ironically because in-store security cameras often focus on the cash till and the customer, film of the customer entering their PIN number is recorded and can be replayed to reveal the exact number entered. Although those who

7 Op. cit. Chapter 1: 1.
8 Op. cit. Chapter 1.

instituted the 'chip and pin' technology did so in order to improve security of credit and debit card use, it is clear that it has generated new, unanticipated forms of insecurity.[9]

Those who operate surveillance technologies are scarcely in a less ambivalent position (Goold, Nellis, this volume). A common view is that the watchers have panoptic powers, that the tools they wield furnish them with total knowledge. Personified in the imagery of Orwell's Big Brother, the watchers are feared and revered. Yet as Smith's contribution to this volume makes clear, they are often far less potent than literary imagery and received wisdom suggest. As Smith observes of CCTV operatives: '[d]isempowerment is . . . compounded both by the operators' structural and hierarchical impotence over other agents within the surveillance web and . . . through their *in*capacity to make a *physical* difference in the mediated action taking place on the screens' (Smith, this volume). Knowledge may be power, but knowing alone is not sufficient to counter the impotence that results from an inability to interact in a determinative way with the technologies under their supposed control. The contradiction between our perception of surveillance operators as empowered watchers and their own sense of incapacity to act upon what they see highlights the contradictory qualities of the human dynamics of security technologies.

Likewise the relationship between technology and those who are supposedly subject to it is more contested than the presumed hierarchical relationship between watchers and watched allows. Given that those subject to technological intervention are often in any event poor and relatively powerless, the effect of technology may be further to disempower, to alienate, to oppress, or to promote social exclusion (Jewkes, this volume). Social sorting is a core activity of many security technologies and Lyon brilliantly observes the terrible and sometimes tragic impact of the categorisation, grading, exclusion and even obliteration of those subject to surveillance (Lyon this volume, see also Neyland).

And yet the top-down structure presumed by the 'sur' in surveillance is by no means inevitable or ubiquitous. The example of amateur photographers discussed by Koskela raises questions about whether the term 'surveillance' suffices to capture what is in play. For whereas surveillance denotes a watching from above, typically by the state, the proliferation and miniaturisation of inexpensive new technologies such as mobile phones makes possible a new genre of *sousveillance* (Mann, Nolan et al 2003) whereby ordinary citizens capture and record incidents in ways that may subvert the authority of officials, police or security guards. The result is, as Koskela observes, that '[t]he differences between the authorities and the public, outsiders and insiders, the

9 See, for example, the BBC ' "Chip and pin" security warning' http://www.news.bbc.co.uk/1/hi/business/4108433.stm and 'Has chip-and-pin failed to foil fraudsters?', *The Guardian* (03.01.2008) http://www.guardian.co.uk/technology/2008/jan/03/hitechcrime.news

controlled and the controllers, have become less clear' (Koskela, this volume). The more recent proliferation of shared internet spaces such as YouTube and Facebook call further into question the hierarchical assumptions of the term of surveillance. By making possible altogether more democratic, non-hierarchical shared spaces or 'commons' where information is readily accessible and refutable by all, these newer technologies signal what has been termed variously the rise of 'co-veillance', 'inverse surveillance' or, most optimistically, 'equiveillance',[10] this latter term suggesting the possibility of a state of equilibrium or social balance as between sur- and sousveillance, between watcher and watched. If equiveillance is increasingly a technological possibility, persisting inequalities in power and resources mean it remains better understood as a utopian aspiration than a present reality. Nonetheless it does suggest that the rigid categories of agent, operator and subject are more fluid than once they were and may even be breaking down altogether.

IRONIES AND INCONGRUITIES OF SECURITY TECHNOLOGIES

Perhaps the most notable collective achievement of the contributions to this volume is to reveal the inherent conflicts and contradictions, inconsistencies and ironies in the pursuit of security through technology. Teasing out these dualities across the chapters reveals some intriguing common characteristics and trends.

Panacea or peril?

Given that technology is often presented as a panacea to the multiple challenges of insecurity, there is a deep irony in the fact that despite avowedly being dedicated to the endgame of security, technological developments instead construct new dimensions of insecurity. As Jewkes and, especially, Dahl's contributions make clear, apparent solutions to security problems are less scientific and less reliable than the claims made for them assert. Part of the explanation lies in their intrinsic fallibilities, but it lies also in the uses to which they are put. For example, the insecurity of DNA science resides as much in the fact that it has been deployed overwhelmingly as an inculpatory technology to affix guilt, despite the fact that in theory it could equally be used to exonerate as to incriminate. The example of DNA usage in the courts illustrates the ways in which technologies are constructed to fit specific political and social ends, and flags up the importance of the context in which they are deployed. The greater the priority given to security in particular

10 Mann, S. (2005) 'Equiveillance: The equilibrium between Sur-veillance and Sous-veillance', http://www.idtrail.org/files/Mann,%20Equiveillance.pdf

locales – the prison or the courtroom, for example – the more we demand of technologies and the less critical we appear to be in our scrutiny of them. Our investment in the court room as a secure place and the confidence we collectively place in the certainty of the conviction does not allow for the acknowledgement of ambiguity. The resultant risk is that what is promised as a panacea becomes a peril: as Dahl reveals in respect of DNA evidence, the security technology turns out to be an unanticipated source of insecurity.

If there is a moral from this tale it is that prudence requires that we treat apparent techno-solutions to security problems with the same caution that the product of any human endeavour deserves. Where information generated by technological innovation is relied upon as a basis for policymaking or, more perilously still, as evidence in court that reliance calls for extreme caution and a willingness continually to test its veracity and accuracy. The example of earlier instances of largely uncritical investment in the reliability of technologies should stand as dire warning of the dangers of doing otherwise. The uncritical faith place in fingerprinting is a prime example here (Cole 2003). Although new technologies are often presented as setting a gold standard of information retrieval or personal identification for security or law enforcement ends, their status as scientific or legally reliable evidence is variable, and the claims made for them are too. Regardless of their capacity for increasing security, insecurities also attach.

Criminologists are only now beginning to document the crimogenic implications of modern technologies (Yar). Computer and internet security and cyber-crime control have received relatively sparse criminological attention despite the fact that the 'informational economy' is a central facet of modern life. Although information technology, and above all the internet, is an undoubted means of human flourishing, it also generates significant new insecurities. As access to information becomes ever more central to economic competitiveness, so isolation from the internet compromises the ability of individuals to function and flourish. Since it is commonly those most socially and economically marginalised that have least ready access (see Jewkes' exploration of prisoners' access to the internet, this volume), so denial of access aggravates pre-existing ontological insecurity. It follows that whilst technologies are potential sources of insecurity, denial of access to these technologies can also foster the insecurities inherent in social and economic exclusion.

Concrete or construct?

One means of instituting a more cautious approach to new technologies lies in the recognition of a second duality, namely between technologies as tangible, material advances and technologies as questionable constructs. Reliance upon science-based technologies to control human behaviour is a central aspect of modernisation. And as Gary Marx has observed, in the field of crime and security 'the search for the illusive silver bullet' (Marx 2001) has placed a special onus upon invention and technological experimentation. Despite the fact that the scientific solutions consequently derived are being applied to social problems, the claims of scientific objectivity and impartiality continue to adhere. Marx further observes: 'Technological controls, presumably being science based, are justified as valid, objective, neutral, universal, consensual and fair' (ibid.).

As many of the contributions to this volume reveal, however high-tech they may be, in practice concrete technological innovations are mediated through human agency. They are influenced by the ambitions of their generators, the cultural mentalities and personal interests of their operators and the demands of their consumers. Moreover, the data they produce is gathered and interpreted for particular political or social ends. Commonly this data (whether generated by satellite tracking systems, CCTV, new computer technologies or biometric identification) tells only a small part of a more complex story that must be pieced together to form a plausible narrative. CCTV images, like biometric science data, are presented as 'realistic evidence depicting objective truth' (Smith, this volume) but may equally be liable to manipulation or even abuse. At best, they may properly better be understood as no more than 'jigsaw pieces in criminal cases' (Dahl, this volume). And Smith wryly observes, how surveillance operators 'fuse together pieces of reality (evidence in a sequential montage (production)' (Smith, this volume) in order to tell particular, and possibly partial, stories.

Tangible technologies thus need to be understood also as social constructs whose meaning is the product of human interpretation and whose significance is inseparable from the uses to which human actors put them. Likewise sources of insecurity may be the product of remarkable reconstructions. So that, as Neyland makes clear, our perception and handling of mundane everyday objects can be radically altered when what was once regarded as ordinary is recast as a potential terrorist threat (Neyland). His analysis recalls Jonathan Simon's earlier brilliant observations on the construction of other everyday objects as security devices. Thus, for example, the mobile phone and the sports utility vehicle (SUV) are marketed, sold and, it would appear, bought not only to serve as means of communication and travel but as perceived sources of safety in an increasingly insecure world (Simon 2007: 201).

Avowed aims and underlying purposes

Closely related to these contested readings of technology as panacea or potential peril, as concrete solution or social construct, is a third dualism: namely the gap between avowed aims and underlying purposes. Security technologies self-evidently have as their declared aim the pursuit of security. And yet as the example of air-safety management systems discussed above made clear, even in the highest security settings measures may have ulterior purposes which are as much about ensuring profitability, averting political fallout should disaster occur, satisfying the demands of particular constituencies or simply pleasing consumers as about improving security per se. The example of the sale of SUVs given above is replete with irony, for as Simon observes: 'The SUV is promoted as a form of security, but that security, to the extent that it is real, is focused only on the occupants of the SUV. To others on the road, even other SUV drivers, SUVs increase the risk of accident', not least because they reduce visibility and encourage aggressive driving (Simon 2002: 145). Even for its users, what is sold as a potent, positively steroidal symbol of security turns out to offer less protection than the sealed capsule that constitutes the average saloon car. The selling of the SUV as a mobile security environment might better be read as a marketing ploy that plays on people's insecurities to augment sales.

This disjuncture between the selling of security technologies and their ulterior purposes applies equally to political salesmanship. As Lyon acutely observes, the history of the ID card abounds with evidence that a device sold by states as a benign classificatory system for making citizens safer has repeatedly resulted in greater insecurity for minority groups such as immigrants and refugees. The official endorsement of the ID card as an administratively efficient means to security conceals a less palatable history of a tool whose function has been to manage populations for reasons entirely other than security. As Lyon observes: 'The new ID cards of today have a long and not always distinguished pedigree that should not be forgotten' (Lyon this volume).

Where security technologies are mediated through professional operators, the picture is complicated further by the fact that the motives of operators may differ from those who generated or instituted the technologies in the first place. In respect of CCTV, for example, the ostensible policy goal of security is undermined by camera operators' more mundane interest in alleviating boredom or seeking titillation (Goold 2004: Chapter 6). Understanding security personnel (whether police, private security guards or camera operators) as 'embodied agents or 'corporeal guardians of control' requires closer attention to the interactions between man and machine so that the uses to which technologies are actually put are never simply assumed (Smith, this volume).

The question of ostensible aims and underlying purposes becomes all the more pressing the more intrusive or injurious the means used. As Halvorsen's contribution makes clear, techniques of torture are justified as necessary in the case of 'ticking bomb' scenarios without sufficient recognition of the fact that these scenarios are thought-experiments rarely mirrored in real life. The overwhelming imperative to avert the hypothetical ticking bomb cannot justify real-life action and the attempt to justify the use of torture techniques must call into question the covert intentions of those who deploy hypotheticals in this way to justify the introduction of otherwise unjustifiable measures (Gross 2004). Although those who seek to deploy insecure technologies may do so for a noble cause (in Dahl's example to secure the conviction of dangerous offenders, in Halvorsen's analysis to avert catastrophic terrorist attack), recognition of their limits ought to act as a sharp brake on so seeking security.

CONCLUSION

In a recent challenging book, Bernard Harcourt observes that the rise of actuarial technique 'is deeply troubling because it demonstrates the influence of technical knowledge on our sense of justice. We have become slaves of our technical advances' (Harcourt 2007: 32). There is indeed a profound irony in the fact that technologies developed in pursuit of our most utopian visions appear instead to generate dystopic outcomes whose perils we now struggle to control. Technologies with the potential to protect also have the capacity to disempower, to alienate, to oppress and to endanger those subject to their use. The securing of one space may be bought at the price of rendering another yet more insecure and the protection of self bought at the cost of greater insecurity for others.

And yet, read together, the contributions to the present volume suggest that this seemingly gloomy scenario is escapable, that there are means of resistance at hand. Their identification lies in the careful observation and analysis of the ways in which technologies of security have been developed and deployed. Micro-analyses of the workings of specific technologies, the relations between man and machine, and the cultural, political and structural contexts in which they operate allow thick description to replace dystopic prophesies. Upon the basis of these the inherent insecurities of security technologies can be laid bare and, by being laid bare, checked. By identifying the limits to security technologies we can perhaps overcome the modernist tendency to techno-credulity and tame the fiend we fear we have inadvertently created. If we cannot entirely master technology, neither need we be its slaves.

References

Aas, K.F. (2007) *Globalisation and Crime*, London: SAGE.

Civil Aviation Authority Safety Regulation Group, *Safety Management Systems for Commercial Air Transport Operations: A guide to implementation*, CAP712, available at http://www.caa.co.uk

Cole, S.A. (2003) *Suspect Identities: A history of fingerprinting and criminal identification*, Boston: Harvard University Press.

Erikson, K.T. (1995) *A New Species of Trouble: The human experience of modern disasters*, New York: W.W. Norton & Co.

Feeney, F. (1985) 'Interdependence as a working concept', *Managing Criminal Justice*, D. Moxon: London, HMSO.

Fisher, E. (2002) 'Precaution, precaution everywhere: Developing a "common understanding" of the precautionary principle in the European Community', *Maastricht Journal of European and Comparative Law*, 9(1): 7–28.

Goold, B.J. (2004) *CCTV and Policing: Public area surveillance and police practices in Britain*, Oxford: Oxford University Press.

Gross, O. (2004) 'Are torture warrants warranted? Pragmatic absolutism and official disobedience', *Minnesota Law Review*, 88: 101–175.

Haggerty, K.D. and Ericson, R.V. (2000) 'The surveillant assemblage', *British Journal of Sociology*, 4(1): 605–622.

Hale, A.R. and de Kroes, J. (1997) 'System in safety', *Safety Science*, 26(1/2): 3–19.

Harcourt, B. (2007) *Against Prediction: Profiling, policing, and punishing in an actuarial age*, Chicago: University of Chicago Press.

Lyon, D. (1994) *The Electronic Eye*, Cambridge: Polity Press.

Mann, S., 'Equiveillance: The equilibrium between Sur-veillance and Sous-veillance', available at http://www.idtrail.org/files/Mann,%20Equiveillance.pdf

Mann, S., Nolan J. et al (2003) 'Sousveillance: Inventing and using wearable computing devices for data collection in surveillance environments', *Surveillance and Society*, 1: 331–355.

Marx, G.T. (2004) 'What's new about the "new surveillance"?', *Surveillance and Society*, 1(1): 9–29.

Marx, G.T. (2001) 'Technology and social control: The search for the illusive silver bullet', in N.J. Smelser and P.B. Balte (eds) *International Encyclopedia of the Social and Behavioural Sciences*, New York: Elsevier.

Marx, G.T. (1989) *Undercover: Police Surveillance in America*, California: University of California Press.

McCahill, M. (2002) *The Surveillance Web: The rise of visual surveillance in an English city*, Cullompton: Willan.

Simon, J. (2007) *Governing Through Crime: How the war on crime transformed American democracy and created a culture of fear*, New York: Oxford University Press.

Simon, J. (2002) 'Guns, crime and governance', *Houston Law Review*, 39(1): 133–148.

Zedner, L. (2008) 'Fixing the future? The pre-emptive turn in criminal justice,' in S. Bronnit, B. McSherry and A. Norrie (eds) *Regulating Deviance: The redirection of criminalisation and the futures of criminal law*, Oxford: Hart Publishing.

Zedner, L. (2006) 'Neither safe nor sound? The perils and possibilities of risk', *Canadian Journal of Criminology and Criminal Justice*, 48(3): 423–434.

Index